GREEN BUILDING THROUGH INTEGRATED DESIGN

McGRAW-HILL'S GREENSOURCE SERIES

Gevorkian
Solar Power in Building Design: The Engineer's Complete Design Resource

GreenSource: The Magazine of Sustainable Design
Emerald Architecture: Case Studies in Green Building

Haselbach
The Engineering Guide to LEED—New Construction: Sustainable Construction for Engineers

Melaver and Mueller (eds.)
The Green Building Bottom Line: The Real Cost of Sustainable Building

Yudelson
Green Building Through Integrated Design

About *GreenSource*
A mainstay in the green building market since 2006, *GreenSource* magazine and GreenSourceMag.com are produced by the editors of McGraw-Hill Construction, in partnership with editors at BuildingGreen, Inc., with support from the United States Green Building Council. *GreenSource* has received numerous awards, including American Business Media's 2008 Neal Award for Best Website and 2007 Neal Award for Best Start-up Publication, and FOLIO magazine's 2007 Ozzie Awards for "Best Design, New Magazine" and "Best Overall Design." Recognized for responding to the needs and demands of the profession, *GreenSource* is a leader in covering noteworthy trends in sustainable design and best practice case studies. Its award-winning content will continue to benefit key specifiers and buyers in the green design and construction industry through the books in the *GreenSource* Series.

About McGraw-Hill Construction
McGraw-Hill Construction, part of The McGraw-Hill Companies (NYSE: MHP), connects people, projects, and products across the design and construction industry. Backed by the power of Dodge, Sweets, *Engineering News-Record* (*ENR*), *Architectural Record*, *GreenSource*, *Constructor*, and regional publications, the company provides information, intelligence, tools, applications, and resources to help customers grow their businesses. McGraw-Hill Construction serves more than 1,000,000 customers within the $4.6 trillion global construction community. For more information, visit www.construction.com.

GREEN BUILDING THROUGH INTEGRATED DESIGN

JERRY YUDELSON, PE, MS, MBA, LEED AP

New York   Chicago   San Francisco   Lisbon   London   Madrid
Mexico City   Milan   New Delhi   San Juan   Seoul
Singapore   Sydney   Toronto

![The McGraw-Hill Companies]

Library of Congress Cataloging-in-Publication Data

Yudelson, Jerry.
    Green building through integrated design / Jerry Yudelson.
        p.   cm.
    Includes index.
    ISBN 978-0-07-154601-0 (alk. paper)
    1. Sustainable buildings—Design and construction. 2. Building—
Methodology. 3. Sustainable design. I. Title.
    TH880.Y635   2009
    721'.0467—dc22   2008030633

1 2 3 4 5 6 7 8 9 0 DOC/DOC 0 1 4 3 2 1 0 9 8

ISBN 978-0-07-154601-0
MHID 0-07-154601-4

Sponsoring Editor: Joy Bramble Oehlkers
Production Supervisor: Pamela A. Pelton
Editing Supervisor: Stephen M. Smith
Project Manager: Somya Rustagi, International Typesetting and Composition
Copy Editor: Nigel Peter O'Brien, International Typesetting and Composition
Proofreader: Anju Panthari, International Typesetting and Composition
Indexer: Broccoli Information Management
Art Director, Cover: Jeff Weeks
Composition: International Typesetting and Composition

Printed and bound by RR Donnelley.

McGraw-Hill books are available at special quantity discounts to use as premiums and sales promotions, or for use in corporate training programs. To contact a special sales representative, please visit the Contact Us page at www.mhprofessional.com.

 The pages within this book were printed on acid-free paper containing 100% postconsumer fiber.

About the Author

Jerry Yudelson, PE, MS, MBA, LEED AP, is the Principal of Yudelson Associates, a green building consultancy based in Tucson, Arizona. He holds engineering degrees from the California Institute of Technology and Harvard University, as well as an MBA (with highest honors) from the University of Oregon, and he is a licensed professional engineer (Oregon). Mr. Yudelson has spent his professional career engaged with energy and environmental issues, and has been involved on a daily basis with the design, construction, and operation of residential and commercial green buildings. He works for architects, developers, builders, and manufacturers to develop sustainable design solutions. His work on design projects involves early-stage consultation, eco-charrette facilitation, and providing LEED expertise and coaching for design teams. He works with developers and building teams to create effective programs for large-scale green projects, as well as with product manufacturers to guide them toward sustainable product marketing and investment opportunities. In addition to this general business and professional background, Mr. Yudelson serves as a LEED national faculty member for the U.S. Green Building Council (USGBC). Since 2001, he has trained more than 3500 building industry professionals in the LEED rating system. He has served on the USGBC's national board of directors and, since 2004, he has chaired the steering committee for the USGBC's annual conference, *Greenbuild*—the largest green building conference in the world. He is the author of *Green Building A to Z: Understanding the Language of Green Building; The Green Building Revolution; Choosing Green: The Homebuyer's Guide to Good Green Homes;* and *Marketing Green Building Services: Strategies for Success.*

CONTENTS

FOREWORD

In the year 2000 when I first began working to introduce green building at Harvard, the most common perception I encountered was that green building was too expensive and that LEED was a costly point-chasing exercise with no value. Things reached their lowest point in 2001 when at one design team meeting, a faculty member acting as the project's client representative likened the belief that green building design could be cost effective to believing that there were elephants in the hallway.

To help transcend these attitudinal barriers, in 2001 I found three building project partners who agreed to pilot LEED at Harvard University. By studying these projects, I was able to trace almost all of the criticisms leveled at LEED to *a range of failures in the design process itself rather than failures intrinsic to LEED.*

For example, the complaint that LEED certification was too expensive turned out to be the result of architects overcharging because they had little experience and were trying to cover their own learning costs and perceived risks. The complaint that there were too many unexpected costs turned out to be the result of change orders that were in turn a result of poor integration of LEED requirements into the building construction documents. The accusation that LEED was a point-chasing exercise turned out to be the result of flawed sequencing of tasks such as the engineer doing the energy modeling after the design was already complete in order to satisfy the LEED documentation requirement, instead of doing it early enough to inform the design.

These pilot projects provided the necessary experiential evidence to prove that green building and the LEED framework in particular did have enormous value if utilized properly. Perhaps most importantly these projects proved to me that cost impact was largely subject to our own ability to properly manage the design process itself and that we needed to stop trying to answer the question "How much will green building and LEED cost us?" and start answering the question "How do we improve the design process to minimize or avoid additional costs for green building and LEED?"

By successfully working to answer this question at Harvard, my team and I have now [Summer 2008] engaged the Harvard community in over 50 LEED projects, most now striving for LEED Gold certification. Utilizing this momentum we were able to work with the extremely decentralized Harvard community to define and adopt a set of comprehensive green building guidelines that includes many key design process requirements, along with a minimum LEED Silver requirement. At the same time I have been working to foster the capacities of both the Harvard community and the building profession that serves it by leading an effort to get everything that we have been learning about the process of green building into a publicly available web resource.*

---

*See www.greencampus.harvard.edu/theresource, accessed July 31, 2008.

Which brings me to why I am so enthusiastic about this book. It is an important resource for anyone who wants to leapfrog years of experiential learning and get right to the heart of effective design process management for green building design. To date very few publications and resources have been focused on the design process and yet in many regards good process management is always the foundation for sustained and successful innovation.

To help get you in the mindset for this process-rich publication, here are my Ten Commandments of Cost-Effective Green Building Design:

1 *Commitment.*   The earlier the commitment is made, the better for everyone. This should be a formal, continuously improved, widely known, and detailed green building commitment for all building projects, integrated into capital project approval processes and related contracts.

2 *Leadership.*   To minimize the risk of business as usual, the client and/or project manager must take an active and ongoing leadership role throughout the project, establishing project-specific environmental performance requirements in pre-design (LEED is ideal for this), challenging, scrutinizing, and pushing the design team at every stage. The client and/or project manager should understand enough about LEED, integrated design, energy modeling, and life-cycle costing to ask the right questions at the right time, a subject this book goes into at length.

3 *Accountability.*   To avoid lost opportunities and unnecessary costs, establish all roles and responsibilities, sequencing and tracking requirements for every environmental performance goal. LEED is ideal for this purpose. Use the LEED score-card to empower the client to participate actively in holding the project team accountable. Utilize LEED's third-party verification process to keep the design team on track with documentation. Work to streamline LEED documentation procedures by paying attention to (and learning from) every project.

4 *Process Management.*   The failure to properly manage tasks at each stage in the design process results in a wide range of missed opportunities and avoidable costs. Each green building performance goal requires a set of tasks to be identified, understood, allocated across the team, sequenced and integrated properly into the design team process. At every stage in the design process, from predesign through to construction and occupancy, there are stage-specific activities that must be completed to maximize innovation and minimize added costs.

For example, many design teams don't include the building operators, or they fail to get any real value from the energy modeling process (because it is done too late to inform the design) or they fail to incorporate a life-cycle costing approach because cost estimations are either done too late and/or fail to include operating costs in the cost model.

5 *Integrated Design.*   Effective integrated design can produce significant design innovations and cost savings. The client and project manager must commit to integrated design and apply constant pressure on the project team to comply. Commitment to the process must be included in all contracts, the selection process and any ongoing team performance evaluation and quality assurance processes.

The right people must be included at the right time (e.g., future building operations staff, the cost estimator, commissioning agent, and controls vendor), and the team must be managed using a collaborative approach to optimize whole building systems rather than isolated components. Well-facilitated design charrettes during conceptual design and schematic design phases are essential.

6 *Energy Modeling.* Energy modeling should go hand-in-hand with the integrated design process and life-cycle costing. Energy modeling must be used at the right phases in the design process, such as schematic design and design development, to evaluate significant design alternatives, inform efforts to optimize building systems, and generate helpful life-cycle-costing data.

7 *Commissioning Plus!* You should expect failures in both the installation and performance of new design strategies and technologies. Beyond making sure that the project team includes a commissioning agent by the end of schematic design, you should undertake an additional effort to test the entire building to ensure that it is performing according to specifications. Projects should include metering, monitoring, and control strategies to support building performance verification and ongoing commissioning for the life of the building. For complex buildings such as laboratories, include the controls vendor by the end of schematic design to integrate the logic of the operating systems into the design. Be sure to train, support, and effectively hand the building over to the operations staff.

8 *Contracts and Specifications.* All green-building-associated process and LEED requirements must be effectively integrated into the owner's project requirements, requests for proposals, all contracts, and all design and construction documents.

9 *Life-Cycle Costing.* The commitment to utilize a life-cycle-costing approach should be made by the client before the project even begins. This commitment should be integrated into all related contracts and specifications. The cost estimator should be brought on board early in the projects, so that costs can be continuously evaluated, including operating cost projections. Energy modeling should be productively utilized to inform operating cost projections, and building operations staff should be engaged to assist in considering operating-cost alternatives. Ensure that a life-cycle-cost perspective is utilized during any value engineering activities.

10 *Continuous Improvement.* For organizations that own more than one building, lessons from every green building project experience should be intensively mined to inform continuous improvement in the building design process and the ready adoption of proven design strategies and technologies. Utilize LEED documentation to support continuous improvement. Where possible, have someone from your organization act as the clearinghouse for project lessons. Invest in deliberate mechanisms to transfer experience from one project to the next. Invest in measurement and verification strategies to evaluate the actual performance of building features.

It is still a challenge to successfully integrate all Ten Commandments into our projects at Harvard, but with every experience we get closer. Harvard's Blackstone Office historic renovation (cover photo) has come the closest. As a direct result of utilizing many of these strategies, the renovation achieved its LEED Platinum certi-fication in 2007 at no added cost to the project. The 40,000-square-feet project was completed on time in 2006 and on budget with a hard cost for construction of $250 per square foot. The client (owner) team did invest a significant amount of their time reviewing and guiding the project, a real cost that was absorbed by non-project budgets. Interestingly, even this investment of additional client time has resulted in the client group developing a range of spin-off campus service offerings such as an owner's acceptance program now offered by the facilities group, which provides building owners at Harvard with additional building systems testing and better training and support for building operations staff.

Today, at Harvard and across the country, the challenge is less about convincing people to do green building, and more about keeping up with the enormous hunger for knowledge and guidance to help design teams achieve the greenest buildings with the least cost impact. To this end I hope you will find this book to be an extremely timely and highly informative resource for addressing critical aspects of the design process as you too strive to make your contribution to the green building movement.

Leith Sharp
Director, Harvard Green Campus Initiative
Cambridge, Massachusetts

PREFACE

I started this book with one important question in mind: how can building teams design, build, and operate commercial and institutional projects that are "truly green"? In particular, how can we deliver buildings that will save at least 50 percent of energy use against standard buildings, that is, those built just to meet local building code and energy code requirements? In my experience, the building design and construction industry is not sufficiently equipped to achieve these goals in most projects. The disparate incentives and rewards, along with the industry's inherent conservatism, make achieving even minor decreases in energy consumption, measured against prevailing standards (currently the ASHRAE 90.1-2007 standard), difficult. The industry's intense focus on minimizing initial costs, coupled with a short-term mentality among building owners and developers, results in the development of many projects that do not make cost-effective investments in energy savings, even when justified using a 5-year or 10-year investment horizon.

Can we achieve these results with current industry approaches to design and construction? Based on personal and professional experience over the past 10 years, I have concluded that answer is a resounding "no." I decided to write this book with the following simple thesis: we must change the way we design and construct our buildings if we're going to have a chance to reduce overall carbon dioxide emissions below 1990 levels, the current Kyoto target. Otherwise, we may have to live with the consequences of a 37 percent increase in U.S. primary energy use between 2000 and 2020, as predicted by many experts. While a strong case can be made for putting energy conservation in existing buildings first, the fact is that most of today's new construction will still be with us 50 years from now, with energy use built into the building fabric and difficult to change. So, it's good to focus significant attention on new building design, construction, and operations.

Can we achieve these high-performance results with design and construction industry's current structure of incentives and methods? I have observed that the design and construction industry, for the most part, is stuck in a linear, risk-averse mode for delivering buildings, with multiple handoffs between the various parties, and many missed opportunities for doing a much better job. The result is buildings that cost more and perform worse than they need to. Conversely, I've observed a few projects that employed an integrated design process that produced buildings that performed better and cost the same as similar projects. After interviewing dozens of architects, engineers, builders, building owners, and developers, I've concluded that we can do a much better job, but we really need a fuller understanding of the integrated design process. This book is an attempt to answer that need.

The objective of all green building efforts is to build high-performance buildings at or close to conventional budgets. I have found that an integrated design process is the best way to realize this goal. There are good examples of LEED Platinum–certified buildings built for little or no additional capital cost, including the building described in the Foreword, Harvard's Blackstone renovation. Another LEED Platinum project, Oregon Health & Science University's Center for Health and Healing, currently the world's largest, was completed in 2006 at a 1 percent cost premium, net of incentives. Through following an integrated design process, Manitoba Hydro's new 690,000-square-foot headquarters in Winnipeg expects to exceed Canada's Model National Energy Code by 60 percent, in a climate with nearly 70°C (126°F) annual temperature swings. As a government building, the design focus was on long-term ownership economics, including enhancing the health and productivity of the workforce, and providing an exemplary sustainable building.

This book abounds with a number of such real-world examples. From them, I've extracted core principles and practices of integrated design, as practiced by leading architects, engineers, builders, developers, and owners. What I discovered is not a simple formula such as combine A and B, and you get C. It's a more complex management task, one that has to be thought about from the beginning of each project, even at project conception: why do we need this building and where are we going to locate it? To make the task more manageable, I've come up with nearly 400 important questions, largely based on the U.S. Green Building Council's Leadership in Energy and Environmental Design (LEED) green building rating system, that you need to ask at each point in the sequence of planning-design-construction-operations.

Green Building Through Integrated Design was written with the commercial and institutional building designer, owner, and builder in mind. I have worked to, first, understand everything I could about green buildings, and, second, report back to important stakeholders on how to make sense out of a field that's growing 50 to 75 percent a year, a growth rate that results in a doubling in size every 12 to 18 months!

I hope that *Green Building Through Integrated Design* will be your guide to greening your next project. This is *not* a book about how to design a green building—there are many fine books on that subject by leading architects—but rather a book about the design and delivery process. I also show you one of the available project management software tools that will help cut the costs of green building projects, and I present the experiences gained by many fine architects and design teams in dozens of successful LEED Platinum projects.

So, grab a cup of shade-grown, organic, fair-trade coffee, put in a skinny squirt of nonfat milk and some natural organic sweetener, kick back, and let me help you find out from the experts how to design and deliver a high-performance building.

Jerry Yudelson

ACKNOWLEDGMENTS

Many thanks to Leith Sharp, director of the Harvard Green Campus Initiative, for generously sharing her experience by writing the Foreword. Thanks also to Paul Shahriari, GreenMind Inc., for contributing the first draft of the chapter on green building project management software. Leith and Paul are two of the bright lights of the green building movement. Thanks to my editors at McGraw-Hill, Cary Sullivan and Joy Bramble Oehlkers, for championing this book. Thanks in addition to everyone who allowed us to interview them for this book, too many to acknowledge individually, including architects, engineers, facility managers, building owners, and developers. Thanks as well to the many architects, architectural photographers, and building owners who generously contributed project photos for the book. Thanks also to Heidi Ziegler-Voll for the illustrations she created specially for this book. Thanks go to Mike Shea and Eric Ridenour, two architects in Portland, Oregon, for their help with the formulation of the first 100 questions for an early version of the "400 Questions."

A special note of thanks goes to my editorial associate, Gretel Hakanson, for conducting the interviews, helping with the research, sourcing all the photos, and making sure that the production was accurate and timely. This is the fifth green building book that we've worked on together and the value of her contribution grows with each project. Thanks to those experts and friends who reviewed the manuscript and offered helpful suggestions and corrections: Anthony Bernheim, Cindy Davis, John Echlin, Stefanie Gerstle, Nathan Good, Steven Kendrick, James Meyer, Margaret Montgomery, Paul Schwer, and Alan Warner.

Thanks also to my wife, Jessica, for indulging the time spent on yet another green building book and for sharing my enthusiasm for green building.

Finally, many, many thanks to the thousands of passionate green building owners, designers, and builders who recognize the need for sustainable design solutions and work daily to implement them.

GREEN BUILDING THROUGH INTEGRATED DESIGN

THE RECIPE FOR SUCCESS IN HIGH-PERFORMANCE PROJECTS

Throughout this book, I'll be showing you examples of high-performance projects, generally LEED Platinum level achievements and telling you how the participants worked together to achieve these outcomes. In this chapter, I'll show examples from several projects and draw some general conclusions from the experiences of architects, owners, engineers, and contractors. The bottom line: it's very difficult to achieve high-level outcomes without some form of integrated design process.

A leading exponent of integrated design is Boston-based architect William G. (Bill) Reed. Reed is widely credited with being one of the original coauthors of the U.S. Green Building Council's Leadership in Energy and Environmental Design (LEED) rating system. For the past several years, he has been beating the drum for the integrative design process as a way to produce not just "green" buildings but buildings and sites that are actually restorative in process and outcome. Reed is a principal with two firms, Regenesis and the Integrative Design Collaborative. He says this about the integrative design process.*

It's very easy to implement the integrative design process but you have to have a design team and clients that are willing to change the nature of their design process. It's not hard, but it's different. This is about change. How easy is this to do? It really depends on how willing people are to begin a change process.

To do different things with sustainability (which is what we're doing in integrative design), we have to do things differently. To do things differently, we have to think about them in a different way. To think about them in a different way, we actually have to be or become different people.

So how easy is that? The practice aspect is relatively the easy part, the change aspect is very hard. The most successful practice process that we've employed to help people to change their process is at the first charrette. We map out a design process

---

* Interview with Bill Reed, February 2008.

that shows how people are going to be integrating and talking among themselves; when they're going to be talking about it; why they're talking; and the interim deliverables and step-by-step analysis they owe to the team. It's a very detailed map. It's not a critical path; it's an integration roadmap. Without that, people will fall right back into their old practice patterns.

We find that by having a road map, a new person [on the project] can just pick up where another person left off. The road map is very detailed. It says who's meeting whom, when and for what purpose. It bundles the analysis process for all major systems: energy, water, materials, habitat, marketing, construction, master planning, architecture, project management, renewable energy. All of those systems are identified as line items and then brought together into the whole system as the design process moves from concept through schematic design. By the way, if you aren't done with these analyses by schematic design, you've likely missed opportunities for the most environmental and cost effective solutions. It can get quite complex but if you don't have it, the team will likely fall back into isolated decision making and overly simplistic dis-integrated ideas.

Integrative design is simply doing research—applied and direct research—and coming together and talking about the discovered opportunities. Then you look for greater systems optimization by questioning all assumptions and go back out and do more detailed analysis and more research, come back together to discuss and compare, and so on. This is a process that requires taking the time to reflect on a deeper purpose, which is the core reason why we are doing a green building, and then letting the genius of the group emerge from that research and questioning assumptions process.

Typically we find three to five charrettes are required for the average building project. Anybody that says they are doing integrative design with one charrette doesn't know what integration requires. One charrette is not an integration process.

Most new practitioners to green building don't know what they don't know. So how can they create a reasonable fee proposal for this work? Even though we maintain that the basic design services cost the same with integrative design, the costs are front-loaded because you're spending a lot more time in pre- and concept design. However, the rest of the project goes much more smoothly because the process is well coordinated.

You can see from Reed's explanation that the integrative design process is really a fundamental challenge to the notion of the "better, faster, and cheaper" approach that characterizes many design projects. Owners just assume that architects and consultants know what they're doing. By contrast, this approach to integrative design recognizes that good work takes good thinking. For many architects, the hardest part of the process might be to resist the urge to take up a pencil and start sketching. Reed's emphasis here is on creating a fully functional team that does investigations and reports back; in this rendition, the process is highly iterative and is based on discovering hidden relationships and possibilities.

Beginning a project with three to five charrettes is obviously a major commitment of time and money for a design team and for a project owner. Reed discusses how these contracts and commitments can be managed effectively.

We do a Part A and a Part B contract. The Part A contract says that we're going to pay the design team members—landscape architects, civil engineers, mechanical engineers, energy modelers, architects, water systems people—for three days of research, for example. We spell out what they're going to be doing. The contract also says that we'll pay them for a two-day charrette. We ask for a proposal for that work and then we ask for a general, ballpark proposal so we can get idea of what they're thinking of for the rest of the work. But we only commit to Part A, that first research phase and charrette. At the first charrette, we go through the goals of the project or at least the general direction of the project. We road map it so that everybody is aligned around what they need to do, why and how many meetings they've got to attend. Then they can go back and write a detailed proposal. The client and the team members now understand what they're proposing and why.

This really gets everybody on the same page, it's fair to everyone and I've never seen anyone resist it once the road map was drawn. If you're bringing together different design team members that you've never worked with before and you don't quite know their capabilities, some pretty important discoveries are made. A few times we've had people that say, "I didn't know this is what I was signing up for. I didn't know this is what green design required. I didn't know I had to do this level of energy modeling." You find out if in fact you have the right people around the table or are missing expertise that may benefit the project's objectives.

Stephen Kieran, of Philadelphia-based Kieran Timberlake Associates, discussed a contrasting approach to a proposed LEED Platinum Sculpture Building and Gallery at Yale University. In this case, there was only a 21-month window to get the project designed and built; there wasn't much time for extensive studies and elaborative iterative design exercises. The solution: pull together a team of very experienced sustainable design experts and have them collaborate from day one, but without a formal team-building program. It helped that the climate engineering consultants, Atelier Ten, had done many previous projects at Yale, so they were familiar with the university's requirements as well as local climatic factors.[*] Kieran described how the project came together so quickly.[†]

When the client came to us, they came with an extraordinary schedule for the project—meaning it was really fast. In hindsight, that actually helped the whole process. They came and asked whether or not we, as team, could deliver the building from programming through occupancy in 21 months, which is less than half of the normal time. That immediately required re-engineering and redesigning the process because they weren't structured to do that. They're a big institution and

---

[*]Interview with Patrick Bellew, Atelier Ten, London, February 2008.
[†]Interview with Stephan Kieran, March 2008.

they needed to make themselves available to make decisions. We immediately engaged in a process design exercise and brought in a team of consultants. First, we decided that we had to meet every other week in New Haven with university representatives who could make decisions at the table with us as we went. All of the principals of the consultant teams, similarly, needed to be at these meetings so that there was always the capacity to make decisions on all fronts in all of the meetings. Second, we mapped out the university's decision-making process and found out when we would need to get in front of the officers and their design advisory committee (for review). It was all rapid fire, literally, with a lot of meetings.

In the bi-weekly meetings, we had in the room: Atelier Ten as environmental consultants; ourselves as architects; John Morrison, partner of CVM, structural engineers; one of the principals of BVH mechanical engineers; the landscape principal at Andropogon; civil engineers from BVH and a variety of specialized consultants along the way that we added as necessary, who would come in and out on a topical basis. This often related to topics such as the development of the curtain wall and the building envelope. We brought in a research team from our office as well, and a number of them participated regularly in those meetings as well as in the development of the environmental strategies. It was integrated not just across our team but across the owner's team as well, which is fairly unique. Normally we don't design with owners. We just bring materials to owners. This was so rapid fire that they were in the room a lot while we were managing the development of this design process.

What happened was pretty interesting. When we got everybody in the room, the structural engineers would start to talk about mechanical systems. The mechanical engineers would start to talk about structural systems. Even the landscape architect got into it—discussing things inside the building. The [disciplinary] boundaries basically evaporated in the discussions.

As an industry, usually we segregate these things. People don't want to sit in a meeting and hear another discipline talk. The mechanical engineers don't want to sit around for a structural coordination meeting, and vice-versa. We didn't have the time for that [conventional approach]. These were half-day meetings; everybody sat in the same room. What transpired was pretty remarkable because they started to comment on each other's work. For instance, we had some of our best criticism about mechanical systems from our structural engineer. It became a real broad vetting process. Ideas bubbled up to the surface and were more thoroughly integrated as a result of having everyone in the room.

For example, the building has a vertical displacement ventilation system [which is placed in large vertical cabinets]. It's the first one Yale has done. It's completely married to the structure of the building. All of the ventilation cabinets, the outlets for this very low-velocity ventilation, are built around all of the structural columns. We did that for the long-term flexibility of the building, [because] the structure is not going anywhere and the displacement ventilation system needs to become a fixture in the building. We didn't want to compromise the flexibility of the building from the owner's perspective; it's basically a loft building.

We married mechanical system locations to structure because the structure wasn't going to move, so it didn't impede the flexibility by putting the two together. Normally that's something the mechanical engineer and structural engineer would hate to do. They don't want to design around each other's work. Normally, they want to be independent. [This approach] worked really well. The engineers coordinated all of the details to make it happen. It's beautiful, I have to say, when you look at it. We have structure and system integrated into the same aesthetic form.

PLATINUM PROJECT PROFILE

Yale University Sculpture Building & Gallery, New Haven, Connecticut

Yale's 51,000-square-foot Sculpture Building & Gallery houses the undergraduate sculpture program of the School of Art. The $52.6 million, three-story building includes a gallery, administrative offices, classrooms, and machine shops. A sophisticated indoor air quality system continuously tests for indoor air pollutants, then flushes and recirculates the air as needed. A unique curtain wall system combined with solar shading reduces heat gain through the glass façade by about 30 percent. Over 90 percent of the gallery's roof is vegetated.*

© Peter Aaron/Esto.

*Yale Office of Sustainability [online], http://www.yale.edu/sustainability/sculpture.htm, accessed April 2008.

Kieran's description highlights that the pressure to perform does bring disparate viewpoints together. The uniqueness of having the owner in the design meetings, with full authority to make decisions on behalf of the university, is something that other teams can learn from. High-performance design needs iterations, but it doesn't need and can't really tolerate, waiting around for some higher-ups to review design decisions (who haven't been part of the process). The role of the architect in persuading different design disciplines to work together is the hidden message in this tale. Beauty comes out of a fully integrated design. There's something inherently pleasing about a human body, for example, that no robot can duplicate, because it's the product of a fully integrated design (intelligent or otherwise!).

The fast-paced project has many side benefits other than forcing an integrated design process. After all, the team still had to have organizing and "process mapping" meetings at the beginning, but it didn't have time for everyone to go back to their offices, get involved with a half-dozen other projects, and then somehow do some analysis just in time for the next design meeting. Kieran said:

> Slower isn't necessarily better artistically or on a performance basis. In some ways forcing people to make decisions—to just be decisive and move [along] almost creates better architecture than some of the hand-holding and angst that can go on when you don't have that type of a fast-moving process.

Another key lesson from this tale is that principals need to stay with the project. For this process to work well, calendars have to be cleared and meetings must be attended by all key players—no excuses! Too often in the design process, the principals delegate meeting attendance to project managers, but still reserve the right to review design decisions, so the entire process is never fully integrated. Although Kieran doesn't explicitly say it, I'm sure the price of admission to this project for the consultants was the commitment to attend the regular biweekly meetings. If you've ever attended project meetings, you know that no principal of a design firm wants to come to a meeting unprepared to make a contribution, so I will assume that the lead architects and engineering consultants were constantly engaged in this project. It turns out that the builders were engaged from the outset as well. Kieran said:

> By the way, the builders, Shawmut Design & Construction, attended all of the biweekly design sessions and provided cost information from the outset. They were a firm that wasn't necessarily noted for building high-performance buildings, but they got into it and developed the whole process along with us. They were pricing it as we went.

Let's look at the engineer's perspective on integrated design and high-performance buildings. Kevin Hydes is an engineer and also currently chairman of the World Green Building Council. He has previously served as chairman of the U.S. Green Building Council and is an honorary member of the Royal Architectural Institute of Canada. Kevin brings a strong cross-disciplinary view to the subject.*

---

*Interview with Kevin Hydes, Stantec Consulting, January 2008.

From an engineer's perspective, one of the challenges with the integrated design process (IDP) is a fundamental lack of appreciation of what architects do. I think what's different between the training of architects and the training of engineers, from what I can tell, is that architects are fundamentally trained in the idea of integration.

Architecture is ambiguous and engineering is factual. That's kind of the way we set up the discussion but integrated design, out of necessity, needs to have ambiguity. If we knew the answer we'd already be doing it every single time. But we don't, so therefore we have to open up to some notion of ambiguity, which is where architects typically feel more comfortable than engineers.

If we believe that we have something to offer as engineers, I have yet to find an architect that isn't interested in sitting down, when the paper is completely blank and the opportunities are endless, to bring together the creative process of the architect and engineer and the other disciplines as well.

Architecture as a profession understands fundamentally that engineering needs to be part of it, that economics needs to be part of it and that the environment implicitly needs to be part of it (even if they are not environmental architects). So there's this notion within architecture that you need to include everything. I think that's why, to a certain extent, architecture has embraced the integrated design process.

There are a few rules that I have: You can't build what you can't draw, you can't measure what you can't draw and you can't cost what you can't draw. One thing about drawing a sketch is that it's universal. When I'm in China, Japan, or Korea, I can draw a sketch and I don't have to learn the language. I think the integrated design process really builds itself around visual tools primarily, with less emphasis on the written word because the word is subject to interpretation, takes longer to create and can be edited. Whereas a diagram, image or snapshot frames a possibility very quickly.

In early conversations around what opportunities a certain building presents, a proactive, knowledgeable, up-to-date engineer with a track record would jump to the front of room and set about sketching an idea. Invariably someone stands up, takes a different color pen and modifies the original idea, improves it, no doubt. And a third person might continue that process. It's the idea of leading by example. It takes courage to throw an idea out. It's all about courage, not brashness; there's a difference. It's certainly not about grandstanding either. If someone is twitching and sitting on their hands, the facilitator has to be skilled enough to recognize that creative energy and hand the pen over to that person.

One of the beliefs about integrated design is the whole theory of innovation. One of the critical aspects of innovation is the ability to be wrong. If you fear being wrong, you'll never innovate. The integrated design process has got to make it safe to be wrong, not just to be right. The ability to be wrong is critical to success because otherwise we're just going to stick within our comfort zone and just extrapolate the ideas that we've already got, which means probably not getting out of the box.

To me, the first rule of engineering is that you have to know the answers (not all of my colleagues would agree with me) before you do the calculation. I believe that we're going to make the type of change that we need to make [in sustainable design], but we have to start by knowing what the answer is. And the answer is net-zero energy. The answer is LEED Platinum. [Knowing where we need to go], then can we use our collective, creative skills and talents to get to the answer. The most important part of IDP is being very clear and focused on what the answer must be and then developing a process to get there.

Let's pause for a moment and examine some of these insights. First of all, integrated design is an iterative process, in which one party or another can take the lead at any given point. Second, it's about finding specific solutions to specific problems or design issues, not about simply fitting previous solutions into a new set of shoes. Third, it needs a clear set of goals, such as net-zero energy or LEED Platinum. As we'll see later in this book, the best design efforts yield these results as a by-product of innovative thinking and a willingness to take calculated risks.

PLATINUM PROJECT PROFILE

Operations Centre, Gulf Islands National Park Reserve, Sidney, British Columbia

Located on the waterfront, the Operations Centre houses the Gulf Islands National Park Reserve operations and administrative staff. The 11,300-square-foot (1050-square-meter), three-story facility was designed to consume 75 percent less energy compared to a similar building with conventional mechanical systems. An ocean water heat pump system provides all the heat requirements of the building. A photovoltaic system supplies 20 percent of the building's total energy needs. Low-flow fixtures, dual-flush toilets, and a 7900-gallon (30,000-liter) storage tank contribute to a reduction in potable water use by over 60 percent.*

Between the architect and the engineer, there are differences in focus; the architect is interested in the visual appearance of a building in addition to how it functions for its intended purpose. The engineer is interested in economy of resources, reliability and control of building systems. There's no inherent conflict between these two viewpoints, but there are widely varying levels of skill and experience that each party brings to the design problem of high-performance buildings. That's why the IDP is so important; it helps bridge levels of skill, acceptance of ambiguity, and risk-tolerance among the different parties on the project team to produce something that none of them individually could bring into being. Another aspect of Hydes' approach is the visual rendering of complex systems. Each sketch triggers conversations between

---

*Lloyd Alter, Gulf Islands Park Operation Centre: LEED Platinum [online], http://www.treehugger.com/files/2006/11/gulf_islands_pa_1.php, November 8, 2006, accessed April 2008. Canada Green Building Council [online], http://my.cagbc.org/green_building_projects/leed_certified_buildings.php?id=41&press=1&draw_column=3:3:2, accessed April 2008.

right brain and left brain; each time someone says, "what if we did it this way?" and takes up the pen to extend the sketch, there is a creative moment in which everyone is brought along to find better solutions.

Finally, there's a hidden gem in Hydes' remarks. To participate fully in the IDP, you have to have a "track record," in other words, you should have successful experience with previous projects that allows you the freedom to be wrong with a given design idea. High-performance building design is an intense process, in most cases using highly trained and experienced practitioners, tight deadlines and limited budgets. There is a lot of pressure to move quickly to find basic solutions, lock in on them and then turn the details over to the rest of the team. The experience of the experts we interviewed for this book is that you have to resist this tendency to close off promising design avenues until you have fully explored the terrain of possibilities. That's why "knowing what you're doing" is so important. There's no time during a design charrette to go research good ideas. You need to bring them to the table and then be creative with using them. "No country for old thinking," would be a good title for a high-performance building design meeting.

So far, we've discussed the architect and the engineer. What about the building owner, the person for whom all this effort is being made? How do building owners approach the issue of high-performance buildings? We spoke with Dr. Douglas Treadway, president of Ohlone College, Newark, California, who presided over the LEED Platinum design of a new community college campus in the San Francisco Bay area. Here's his take on the process.*

My role in designing the new campus was maybe a little bit more hands-on than the normal role of a college president. Prior to my coming on board, they had the funds and had finalized their plans, but I was able to convince the board to take a brief hiatus and re-examine their plans. That was the first thing that I did. I also pulled back the time frame so we could do a green building, because we weren't planning to do one. They had some vague references but no specific vision of what we were actually going to do. [The time frame was pushed back only about six to eight months.]

The goals were determined after a series of planning retreats, visioning exercises, interviews and research within the Bay Area to determine the feasibility of certain approaches to green building. We then tied that into the vision of the new campus, which also had not been targeted. It was going to be a general college, and then we changed it to being a health sciences and technology college. We then had a different rationale for our green building, because the nature of the institution's mission had changed. Repurposing the building from a general college campus to a thematic health science and technology campus was really important in the early design because that drove everything afterward.

We were in discussion with the team about the project goals but we also had our own independent design criteria. The criteria didn't make up the actual physical design

---

*Interview with Dr. Douglas Treadway, Ohlone College, March 2008.

so much as they were the principles and assumptions that would govern the design. The college community and some of the employers in the area developed the goals during a set of retreats and visioning sessions. We gave the architect eight to ten planning concepts. They gave those a voice and put them into a document. The nearby San Francisco Bay estuary became a planning paradigm for the architect's work. They were quite creative in taking what we did and putting it into architectural terms.

It's been a blended process because we didn't have the contractors involved early on. We used a CM-at-Risk process and used Turner Construction Company, who is very experienced with green building, although two of the systems we put in, they had never done before. So they learned too. My role at that point was kind of like the conductor of the symphony. I was very much involved in making sure this project was integrated with our other college campus and with our planning.

In this whole period of continuous iterations, at every decision point, we asked, "Do we cut back? Do we sacrifice the solar? What do we do?" There was a major decision point when we were getting the bids where I said, "I'll go out and raise the $10 million for the furnishings, don't cut the building back any more and don't take any of the sustainability elements out. I'll find matching money." So I went out on a limb during the latter part of the project, because it did get a bit out of our reach. Because we had all of these partners on with us early on, they came back and said, "We'll help you raise the money." And we did.

Number one, the most important key to reaching Platinum certification was the partnership between the architect and the builder. We didn't start out with green [certification] being a goal. However, they were very motivated to push everything and come back to us and offer suggestions. They would ask us, "How about if you did this? Could you do this and would you see the value of it, not just for LEED but in itself?" We helped make a lot of decisions along the way that we thought were good decisions and ones we wanted to make anyway. We didn't just look for a particular standard but they were really aggressive about it. We had a board policy that required LEED certification of all of our buildings. So certification was a must, but we didn't think we'd get to where we got [Platinum].

On the rationale for integrated design using the LEED rating system, Treadway says:

My advice is to use the LEED criteria, not so much to make a certain mark, but because they are very sensible; they really relate to the quality and health of the learning environment as well as to the long-term affordability. Ongoing maintenance costs are harder for most of us to come up with than the initial capital costs. [Considering] "total cost of ownership" is also an environmental principle, and LEED helps you think that way.

Treadway also spoke to the need to handle rapidly rising construction costs by rethinking the building and designing a smaller project, to stay within the original budget.

[You should] also be willing to sacrifice quantity to get more in quality. Most of our college buildings only operate at about 40 percent annual occupancy anyway. Use the building more, maybe have a little less of it, but make sure it's a good quality, healthy building. Sustainability includes all of that. It isn't just energy savings; it's the ecology of the building and its learning value.

PLATINUM PROJECT PROFILE

Newark Center for Health Sciences and Technology, Ohlone College, Newark, California

The Newark Center for Health Sciences and Technology houses Ohlone College's Newark campus, which serves the San Francisco East Bay region. The 135,000-square-foot facility has the capacity for 3500 students. The rooftop photovoltaic system will supply 42 percent of the building's energy needs. Geothermal ground coils (geoexchange systems) and enthalpy energy recovery wheels contribute to a 25 percent improvement in energy performance.*

Photo courtesy of Lou Galiano, Alfa Tech Cambridge Group.

Turner Construction was the CM-at-risk for the Ohlone College project. We interviewed Michael Deane, operations manager for sustainable construction at Turner for his perspective on integrated design and the contractor's role.†

---

*Ohlone College [online], http://www.ohlone.edu/org/newark/core/leed.html, accessed April 2008. Catherine Radwan, Environmentally Sustainable Campus to Earn LEED Platinum Certification, January 28, 2008, accessed April 2008.
†Interview with Michael Deane, February 2008.

[At Turner], because of the number of green projects we have in our portfolio, we've done almost everything at least once, including seven Platinum projects. We just completed the first Platinum-certified high-rise residential building. We have completed several LEED Gold projects that have come in at or less than two percent more than a standard budget.

If we use the integrated design model we, the builders, can provide real value in a focused discussion. We can add a lot of information that will inform the design choices. If we're not at the table, those things don't always come up. Therefore, the finished product, by way of the design, may not be as good as it could have been if it had the benefit of the builder's perspective.

My advice for other contractors is simple: Speak up for a seat at the table. Make sure that the rest of the team understands the value that you bring to the table. Then when you sit down at the table, you better have done your homework and know what you're talking about.

As more projects start to move toward zero net energy and zero waste solutions, the architectural and engineering systems will need to become more adventurous; without active participation from the general contractor and key subcontractors such as the mechanical, electrical and controls contractors, these projects are not likely to work as well as they could. Since the general contractor spends more than 90 percent of the project budget in a typical building program, integrated design without the contractor's active participation is likely to yield suboptimal results.

If so many people are willing, able, and interested in designing and delivering high-performance projects, what then are the real barriers to high performance buildings? There must be some obstacles, otherwise there would be no need for this book. Dan Nall is senior vice president and director of advanced technology for Flack + Kurtz, one of America's premier green building engineering firms. Here's his take on this issue:*

What is the chief impediment to creating sustainable or high-performance buildings? It's a failure of will—failure of will by the owner and failure of will by the principal designers. You've got to just keep in there; when faced with the "Oh my God, it'll cost a fortune," or "This will never work," or what have you, you just have to keep in there plugging away to do what you have to do to make it work. You read that in all of the self-help books about anything from losing weight to becoming a millionaire, but it is true. In order to do a high-performance building, you have to be working with the right team so that all of the principal members are united in the cause, the cause of doing a superlative building. Once that's there and the trust is there and everybody is united in this endeavor, you just have to keep working at it together and searching for ways to overcome those hurdles which inevitably will get in your path.

It's really that simple and, in terms of human relationships, that complex. The very act of doing something different and more challenging than conventional practice calls

---

*Interview with Dan Nall, March 2008.

for an extraordinary exercise of will as well as high-level professional skills and judgment. As we'll see in this book, every high-performance building reaches some point of insurmountable obstacle, whether technical, functional, or financial, that has to be overcome by the collective will of the building team.

Let's take one look back at the practice of integrative design as expressed by Bill Reed. He is adamant that the LEED system is for keeping track of results, not about dictating design decisions; he says, "don't count LEED points; indoor environmental quality, energy use, water, habitat and site issues are all connected—it's a whole system. You can't achieve a quality environment without looking at all of those areas....We rarely score LEED points until the third or fourth charrette. What we do is have people think systematically [about all these things that are interrelated]". Finally, Reed says, "do it all in predesign; you should be getting all of these answers before you're even starting to design the building."

So, let's close this chapter with Reed's main points for designing and delivering high-performance buildings:

1 Don't count LEED points.
2 Think systematically.
3 Do it all in predesign.
4 Practice the "4 Es": Engage everyone, early, with every issue.

PLATINUM PROJECT PROFILE

Alberici Office Headquarters, St. Louis, Missouri

Completed in December 2004, this 109,000-square-foot office building serves as the headquarters for Alberici Corporation, a St. Louis-based construction company.

Photography by Debbie Franke.

The total project cost was $20,100,000 and the payback period is estimated to be 7.5 years. The facility was designed to exceed minimum energy efficiency requirements by 60 percent. Onsite renewable sources generate 17 percent of the building's required energy. A 65-kilowatt wind turbine provides 20 percent of the building's electrical needs and solar panels preheat water for occupant use. A 30,900-gallon cistern collects rainwater from the roof. The harvested rainwater is used in the cooling tower as well as for 100 percent of sewage conveyance. An onsite retention pond captures cistern overflow and is home to a living ecosystem for native flora and fauna.*

*U.S. Green Building Council [online], http://leedcasestudies.usgbc.org/overview.cfm?ProjectID=662, accessed April 2008.

GREEN BUILDINGS TODAY

Green buildings and sustainable design have been major movements in the design, development and construction industry since about 2000, with an accelerating interest since 2005, as shown in Fig. 2.1. Here we see the growth of green buildings, in terms of cumulative LEED project registrations and certifications, both increasing 75 percent in 2007 alone, and 65 percent and 77 percent, respectively, in 2006 versus the prior year. In this respect, the acceptance and practice of green building design, after growing steadily from 2000 through 2005, began accelerating in 2006 and 2007.

However, a large majority of such projects are still at a basic level of green design, as shown by the number of LEED Certified and Silver projects, as a percentage of the total. By the end of the first quarter of 2008, total LEED for New Construction and Major Renovations (LEED-NC) project certifications (U.S. projects only) numbered 1015 (including the four major systems—LEED for New Construction and Major Renovations, LEED for Core and Shell, LEED for Commercial Interiors, and LEED for Existing Buildings: Operations and Maintenance, total certifications were 1405). Table 2.1 shows the relative percentage of high-performance LEED-NC certifications (Gold and Platinum) represent about 32 percent of the total, with Platinum representing some 50 U.S. projects, or 5 percent of the total. Based this analysis, I decided to focus this book primarily on the Platinum projects, since they represent the highest attainment level in the LEED system and are still relatively rare; barely one project out of 20 makes it into the LEED stratosphere.

As of Summer 2008, the largest LEED Platinum new construction project was the Oregon Health and Science University's (OHSU's) Center for Health and Healing in Portland, Oregon, developed by Gerding Edlen Development as a building to suit for the university. This project contains 412,000 square feet of gross floor area in a 16-story mixed-use medical facility. Completed in the fall of 2006, the OHSU project contains a 300-kW onsite microturbine plant, 60-kW of building integrated photovoltaics (BIPV), a 6000-square-foot site-built solar air heater and numerous energy-efficiency

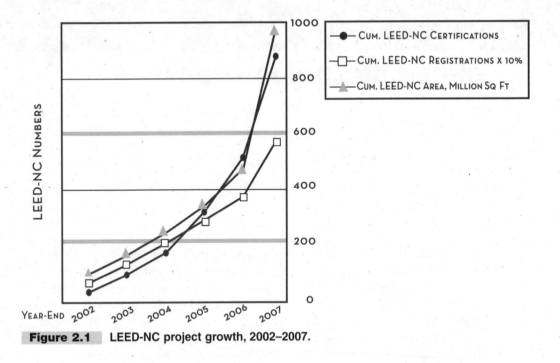

Figure 2.1 LEED-NC project growth, 2002–2007.

measures. The project serves up to 3000 people per day* and treats, then recycles all of the building sewage with an onsite treatment plant. Overall, the project expects to save 60 percent of the energy use of a conventional building and about 56 percent of the water use. It was built for a reported one percent net cost premium over a conventional building.

TABLE 2.1 LEED-NC CERTIFICATIONS BY ATTAINMENT LEVEL, APRIL 2008*		
LEVEL	NUMBER OF CERTIFIED PROJECTS	PERCENTAGE OF TOTAL CERTIFICATIONS
Certified	362	36
Silver	331	33
Gold	272	27
Platinum	50	5
Total	**1015**	**100**

*U.S. Green Building Council data furnished to the author.

*Personal communication, Andy Frichtl, Interface Engineering, February 2008.

PLATINUM PROJECT PROFILE

Center for Health and Healing, Portland, Oregon

Located in Portland's revitalized 38-acre South Waterfront district, the Oregon Health & Science University's (OHSU) Center for Health & Healing includes office space, more than 200 exam rooms, outpatient surgery facilities, educational space, research labs, and a health club and pool. Completed in 2006, the 16-story, 412,000-square-foot building cost $145 million. An onsite micro-turbine plant supplies 30 percent of the building's electrical needs and nearly all of the hot water. All of the wastewater is treated onsite and used for toilet flushing and landscaping, reducing water use by 56 percent. In addition to a 60-kW photovoltaic array mounted on sunshades, the top two stories of the south façade double as a solar heat absorber. Warm air collected behind the 6000-square-foot solar air collector is used to provide hot water for the building. Chilled beams, used in places to supplement traditional air-conditioning, reduce energy use by 20 to 30 percent.*

Courtesy of Gerding Edlen.

---

*Russell Boniface, OHSU Center, First Medical Facility to Win LEED Platinum Award, AIArchitect, volume 14, March 30, 2007 [online], accessed April 2008.

Figure 2.2 Designed by Cook + Fox Architects, with abundant daylighting and use of local materials, the $1 million Bank of America Tower at One Bryant Park in New York City is aiming for a LEED-NC Platinum certification. The project begins move-in during the spring of 2008. *©Gunther Intelmann Photography, Inc.*

What will likely become the world's largest LEED Platinum–certified building, the Bank of America Tower at One Bryant Park in New York City, scheduled to begin occupancy in 2008, provides a new standard for high-rise green buildings worldwide. Shown in Fig. 2.2, this 52-story high-rise commercial building in the heart of New York demonstrates a commitment to sustainable design, as well as investment in the future by the building's owners, Bank of America and the Durst Organization. The project involves 2.1 million square feet of office space, onsite water recycling and an onsite cogeneration plant with ice storage for peak-load reductions. Architectural features include a high-performance glass curtain wall and a crystalline glass look; the building's faceted crystal form allows more sunlight to hit the street, and the reflections off the building will capture the changing angles of the sun. It also reduces wind loads on the building.

The project architects, Cook + Fox, are great supporters of an integrated design process and employed it in this project. According to architect Robert Fox, the keys to integrated design and eco-charrettes for such projects (and which will be found in many of the other projects) are*

*Interview with Robert Fox, Cook + Fox Architects, January 2008.

1 Get "buy in" at the top, including not just the client but also the top management of each important player, including all the consultants.

2 For the initial (and subsequent) design charrettes, engage a competent outside facilitator and give them the authority and flexibility to change the agenda as needed, to ensure results.

3 Make sure that the design charrette agenda is the right one and get everyone to agree to it ahead of time; schedule a multiple-day charrette, with time for social interaction.

4 The charrette must be off-site, with no distractions; there must be time for each key player to respond to the evolution of the design during the charrette.

5 Hold a retreat at the end of each project phase when the players' roles and responsibilities change, to review progress and integrate new people into the project team.

Summarized in another way, we can see that creative, high-performance design takes commitment, creativity, and focus; it also involves people who need to get to know each other, as part of a new project team, at the outset of a project. The process also needs to allow for serendipitous moments, spontaneous events that bring about the "aha" realization, when everything seems to jell for all participants, and the project design moves forward with greater ease and effectiveness.

High-Performance Building Characteristics

Let's get more specific about what we actually mean by the term "green building" or "high-performance building." A green building is one that considers and then reduces its impact on the environment and human health. A green building uses considerably less energy and water than a conventional building, has fewer site impacts and generally higher levels of indoor air quality. It also accounts for some measure of the life-cycle impact of building materials, furniture and furnishings. These benefits result from better site development practices; design and construction choices; and the cumulative effects of operation, maintenance, removal, and possible reuse of building materials and systems.

In the United States and Canada, a green building is generally considered to be one certified by the LEED green building rating system of the U.S. Green Building Council (USGBC) or Canada Green Building Council (CaGBC). More than 99 percent of the "certified" green buildings in both countries come from this system.* For the

---

*Author's analysis, based on reported certifications for new construction projects at the end of 2007. At that time, LEED had certified about 1000 projects, Green Globes about 10, or 1% of the total.

purposes of this book, I will be using "high-performance" buildings to designate those that achieved a Gold- or Platinum-level certification from the U.S. or Canadian LEED systems. This is not entirely a fair choice, because there are some excellent green buildings that only achieved a LEED Silver status; however, increasingly, one must ask that a high-performance building achieve LEED Gold or Platinum. There might even be examples of Zero Net Energy buildings that one would consider "high-performance" from an energy efficiency standpoint that don't achieve a Gold or Platinum level designation, but in my experience, it's highly unlikely.

The essence of LEED is that it is a point-based rating system that allows vastly different green building attributes to be compared with one resulting aggregate score. This gives it the ability to label a wide variety of buildings and to rate very dissimilar approaches to sustainable design with one composite score. Many will (and do) argue with the relative importance of various categories of concern, but this rating system will soon be 10 years old; it has stood the test of time through continuous improvement and openness to intelligent modification and innovation. Because it is point-based, LEED also appeals to the competitive spirit in the North American psyche, which values winning very highly and which associates getting "more points" with winning.

Interestingly, LEED is being seen around the world as a good example of a rating system that is practical to use, but which still signals achievements in sustainable design, construction, and operations far above the norm for most buildings.* Over the years, LEED has continually tweaked its credits to stay at the leading edge of green building design. In 2009, LEED will undergo a major adjustment to increase its flexibility, while maintaining its rigor and credibility, through the adoption of *LEED v3* (or LEED 2009). This new approach is a system that will give design and construction teams far more flexibility in picking which credits to pursue in a given geographic region or for meeting client preferences that might, for example, value energy savings or water savings much more than the current system.

In September 2006, the U.S. General Services Administration reported to Congress that it would use only the LEED system for assessing the government's own projects.[†] The U.S. Army planned to adopt LEED in 2008, to replace its own homegrown "Spirit" rating systems. Therefore, in the commercial and institutional arena, if a project is not rated and certified by an independent third-party with an open process for creating and maintaining a rating system, it can't really be called a green building, since there's no other standard definition.

If someone tells you they are "following LEED" but not bothering to apply for certification of the final building, you should rightly wonder if they would really achieve the results they claim. If they say they are doing "sustainable design," you

---

*For example, a similar approach is being pursued by the German Green Building Council, Deutsche Gesellschaft für nachhaltiges Bauen, which was formed in 2007, www.dgnb.de, accessed July 31, 2008.

[†]General Services Administration, Report to Congress, September 2006 [online], https://www.usgbc.org/ShowFile.aspx?DocumentID=1916, accessed March 6, 2007.

have a right to ask, "Against what standard are you measuring your design, and how are you going to prove it?" By following the LEED guidelines, but not doing the paperwork, the practical effect is that there's a lot less commitment to the final outcome.

COMMERCIAL AND INSTITUTIONAL BUILDINGS

A green building uses design and construction practices that significantly reduce or eliminate the negative impact of buildings on the environment and occupants. In the LEED system, these practices include building location, water and energy use, environmentally preferable purchasing and waste management activities, improved indoor environmental quality and a "continuous improvement" approach to green building innovations. Though owned by the USGBC, the LEED rating system is a publicly available document;* it has an extensive committee structure charged with keeping it current and improving it over time. The current iteration is known as LEED version 2.2. The LEED v3 system (expected in 2009) plans to have greater flexibility for building teams to consider regional issues, a stronger focus on life-cycle assessment and a better way to handle alternative approaches to designing "low carbon" buildings.

Table 2.2 shows the six major categories in the LEED-NC rating system for new and renovated commercial and institutional buildings, mid-rise and high-rise residential towers. At first thought, many people think of a green building as one that primarily uses less energy and possibly uses recycled-content materials. Looking at the entire LEED rating system, one can see that the categories of concern are much broader and more comprehensive than just saving energy. Practitioners in the design, development, and construction industry in the United States and Canada have embraced this system over all other competitors. In this respect, it's fairly easy to say, "It isn't a green building (in the United States and Canada) if it's not LEED certified."

From the standpoint of commercial buildings, the LEED rating system is heavily weighted toward saving energy, saving water, and providing higher levels of indoor environmental quality. However, it's virtually impossible to get a LEED Gold or Platinum rating without paying close attention to the sustainable site criteria and to the need to conserve materials and resources. In this sense, I find that LEED Gold and Platinum projects are fairly balanced across all five main groups of environmental attributes. For example, Table 2.3 shows the LEED scores by category for the two completed buildings at the Arizona State University Biodesign Institute in Tempe, Arizona. You can easily see that the main difference between the two projects is the very high level of energy savings in the Platinum project, more than 50 percent.[†]

---

*U.S. Green Building Council [online], www.usgbc.org/leed, accessed July 31, 2008.
[†]David Sokol, "Crossing Boundaries," Greensource magazine, January 2008, p. 72.

TABLE 2.2 LEED-NC SYSTEM CATEGORIES OF CONCERN*

CATEGORY	TOTAL POINTS	ISSUES EVALUATED BY THE LEED-NC SYSTEM
1. Sustainable sites	14	Avoiding sensitive sites; promoting urban infill; locating to facilitate use of public transportation; reducing site impacts of construction; creating open space; enhanced stormwater management; lowering the urban heat island effect and controlling light pollution.
2. Water efficiency	5	Encouraging water conservation in landscape irrigation and building fixtures; promoting wastewater reuse from onsite sewage treatment.
3. Energy use reduction, green power, and atmospheric protection	17	Energy-conservation; using renewable energy systems; building commissioning; reduced use of ozone-depleting chemicals in HVAC systems; energy monitoring; and green power use.
4. Materials and resource conservation	13	Use of existing buildings; facilitating construction waste recycling; use of salvaged materials, recycled-content materials, regionally produced materials, agricultural-based materials and certified wood products.
5. Indoor environmental quality	15	Improved ventilation and indoor air quality; use of nontoxic finishes and furniture; green housekeeping; daylighting and views to the outdoors; thermal comfort; and individual control of lighting and HVAC systems.
6. Innovation and design process	5	Exemplary performance in exceeding LEED standards; use of innovative approaches to green design and operations.

*LEED 2009 will make changes to the total points and percentage weightings shown here.

TABLE 2.3 BIODESIGN INSTITUTE AT ARIZONA STATE UNIVERSITY. LEED CATEGORY SCORES OUT OF TOTAL AVAILABLE POINTS

CATEGORY	BUILDING A (GOLD)	BUILDING B (PLATINUM)
Sustainable Sites	12/14	12/14
Water Efficiency	4/5	4/5
Energy and Atmosphere	6/17	15/17
Materials and Resources	3/13	5/13
Indoor Environmental Quality	10/15	11/15
Innovation/Design Process	5/5	5/5
Total Points	**40/69**	**52/69**

WHO BUILDS HIGH-PERFORMANCE BUILDINGS?

I analyzed the first 1015 LEED-NC Platinum and Gold buildings, with the results shown in Table 2.4. (I left LEED-CS projects out of this analysis because they are almost all done by the private sector.) The nonprofit or government sector builds more than 75 percent of LEED Platinum buildings and more than 65 percent of LEED Gold buildings. In other words, LEED Platinum projects are still highly institutional in nature, and you'll see that from presentation of many of the LEED Platinum projects in this book.

TABLE 2.4 LEED-NC PROJECTS BY OWNER AND CERTIFICATION LEVEL

TYPE OF OWNER	PLATINUM	PERCENT	GOLD	SILVER	CERTIFIED
For-profit organization	11	22%	94	115	114
Nonprofit Org.	23	46%	59	58	68
Local Govt.	5	10%	48	69	65
State Govt.	5	10%	27	35	44
Other	3	6%	29	27	41
Federal Govt.	1	2%	9	19	26
Individual	2	4%	6	8	4
Total	**50**	**100%**	**272**	**331**	**362**

The LEED Rating Systems

LEED rates all buildings across five major categories of concern, using key environmental attributes in each category. LEED collects and incorporates a wide variety of "best practices" across many disciplines including architecture, engineering, interior design, landscape architecture, and construction. It is a mixture of performance standards (e.g., save 20 percent of the energy use of a typical building) and prescriptive standards (e.g., use paints with less than 50 grams per liter of volatile organic compounds), but leans more toward the performance approach. In other words, LEED believes that best practices are better shown by measuring results (outcomes) not by prescribing efforts alone (inputs).

Each LEED rating system (see Table 2.5) has a different number of total points, so scores can only be compared within each system; however, the method for rewarding achievement is identical, so that a LEED Gold project for New Construction represents in some way the same level of achievement (and degree of difficulty) as a LEED Gold project for Commercial Interiors (tenant improvements). Figure 2.3 shows how the LEED-NC rating system splits points into the five major categories of concern.

TABLE 2.5 THE FOUR MAJOR LEED RATING SYSTEMS FOR LARGE BUILDINGS, SPRING 2008

RATING SYSTEM	TYPE OF PROJECT	PERCENTAGE OF TOTAL REGISTRATIONS*	PERCENTAGE OF TOTAL CERTIFICATIONS
LEED for New Construction (LEED-NC)	New buildings and major renovations; housing more than four stories	66.0	74.0
LEED for Commercial Interiors (LEED-CI)	Tenant improvements and remodels that do not involve building shell and structure	10.3	16.3
LEED for Core and Shell (LEED-CS)	Buildings in which the developer or owner controls less than 50% of tenant improvements	13.7	4.5
LEED for Existing Buildings (LEED-EB)	Ongoing building operations, including purchasing policies	10.0	5.2

*Data for LEED registrations and certifications are from the USGBC, furnished to the author, end of March 2008.

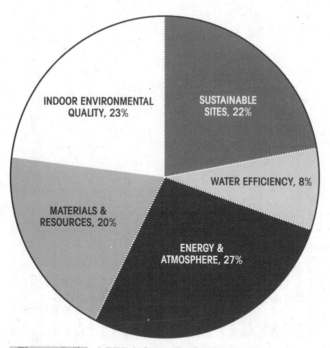

Figure 2.3   LEED-NC credit distribution.

LEED project attainment levels are rewarded as follows:

Certified   >40% of the 64 "core" points in the system

Silver   >50% of the core points

Gold   >60% of the core points

Platinum   >80% of the core points

The LEED rating system is a form of an "eco-label" that describes the environmental attributes of the project, similar to the nutrition labels on food. Prior to the advent of LEED, there was no labeling of buildings other than for their energy use, as found in the federal government's ENERGY STAR® program.* Solely presenting a building's energy use compared with all other buildings of the same type in a given region, gives an incomplete picture of a building's overall environmental impact. The LEED scorecard (shown in Fig. 2.4 for the largest LEED Platinum building in the world) shows a project's achievement in each credit category and allows you to quickly assess the sustainable strategies used by the building team.

The irony here is that a $20 million building that is not LEED-certified has less labeling than a $2 box of animal crackers, in terms of its "nutritional" benefits (energy use,

---

*Energy Star program [online], www.energystar.gov, accessed July 31, 2008.

LEED Scorecard

OHSU Center for Health & Healing 2.21.07

Yes	No	Total Project Score
55	14	Certified 26 to 32 points · Silver 33 to 38 points · Gold 39 to 51 points · Platinum 52 or more points

Sustainable Sites

Yes	No		Credit
Y		Prereq 1	**Erosion & Sedimentation Control**
1		Credit 1	**Site Selection**
1		Credit 2	**Development Density**
1		Credit 3	**Brownfield Redevelopment**
1		Credit 4.1	**Alternative Transportation,** Public Transportation Access
1		Credit 4.2	**Alternative Transportation,** Bicycle Storage & Changing Rooms
1		Credit 4.3	**Alternative Transportation,** Alternative Fuel Refueling Stations
	1	Credit 4.4	**Alternative Transportation,** Parking Capacity
1		Credit 5.1	**Reduced Site Disturbance,** Protect or Restore Open Space
1		Credit 5.2	**Reduced Site Disturbance,** Development Footprint
1		Credit 6.1	**Stormwater Management,** Rate and Quantity
1		Credit 6.2	**Stormwater Management,** Treatment
1		Credit 7.1	**Reduce Heat Islands,** Non-Roof
1		Credit 7.2	**Reduce Heat Islands,** Roof
1		Credit 8	**Light Pollution Reduction**

Section total: Yes 13, No 1

Water Efficiency

Yes	No		Credit
1		Credit 1.1	**Water Efficient Landscaping,** Reduce by 50%
1		Credit 1.2	**Water Efficient Landscaping,** No Potable Use or No Irrigation
1		Credit 2	**Innovative Wastewater Technologies**
1		Credit 3.1	**Water Use Reduction,** 20% Reduction
1		Credit 3.2	**Water Use Reduction,** 30% Reduction

Section total: Yes 5, No 0

Energy & Atmosphere

Yes	No		Credit
Y		Prereq 1	**Fundamental Building Systems Commissioning**
Y		Prereq 2	**Minimum Energy Performance**
Y		Prereq 3	**CFC Reduction in HVAC&R Equipment**
2		Credit 1.1	**Optimize Energy Performance,** 20% New / 10% Existing
2		Credit 1.2	**Optimize Energy Performance,** 30% New / 20% Existing
2		Credit 1.3	**Optimize Energy Performance,** 40% New / 30% Existing
2		Credit 1.4	**Optimize Energy Performance,** 50% New / 40% Existing
2		Credit 1.5	**Optimize Energy Performance,** 60% New / 50% Existing
1		Credit 2.1	**Renewable Energy,** 5%
1		Credit 2.2	**Renewable Energy,** 10%
1		Credit 2.3	**Renewable Energy,** 20%
1		Credit 3	**Additional Commissioning**
1		Credit 4	**Ozone Depletion**
1		Credit 5	**Measurement & Verification**
1		Credit 6	**Green Power**

Section total: Yes 14, No 3

Materials & Resources

Yes	No		Credit
Y		Prereq 1	**Storage & Collection of Recyclables**
	1	Credit 1.1	**Building Reuse,** Maintain 75% of Existing Shell
	1	Credit 1.2	**Building Reuse,** Maintain 100% of Existing Shell
	1	Credit 1.3	**Building Reuse,** Maintain 100% Shell & 50% Non-Shell
1		Credit 2.1	**Construction Waste Management,** Divert 50%
1		Credit 2.2	**Construction Waste Management,** Divert 75%
	1	Credit 3.1	**Resource Reuse,** Specify 5%
1		Credit 3.2	**Resource Reuse,** Specify 10%
1		Credit 4.1	**Recycled Content,** 5% (POST-CONSUMER + 1/2 POST-INDUSTRIAL)
1		Credit 4.2	**Recycled Content,** 10% (POST-CONSUMER + 1/2 POST-INDUSTRIAL)
1		Credit 5.1	**Regional Materials,** 20% Manufactured Locally
1		Credit 5.2	**Regional Materials,** of 20% Above, 50% Harvested Locally
1		Credit 6	**Rapidly Renewable Materials**
1		Credit 7	**Certified Wood**

Section total: Yes 8, No 5

Indoor Environmental Quality

Yes	No		Credit
Y		Prereq 1	**Minimum IAQ Performance**
Y		Prereq 2	**Environmental Tobacco Smoke (ETS) Control**
1		Credit 1	**Carbon Dioxide (CO$_2$) Monitoring**
	1	Credit 2	**Increase Ventilation Effectiveness**
1		Credit 3.1	**Construction IAQ Management Plan,** During Construction
1		Credit 3.2	**Construction IAQ Management Plan,** Before Occupancy
1		Credit 4.1	**Low-Emitting Materials,** Adhesives & Sealants
1		Credit 4.2	**Low-Emitting Materials,** Paints
1		Credit 4.3	**Low-Emitting Materials,** Carpet
1		Credit 4.4	**Low-Emitting Materials,** Composite Wood
1		Credit 5	**Indoor Chemical & Pollutant Source Control**
1		Credit 6.1	**Controllability of Systems,** Perimeter
1		Credit 6.2	**Controllability of Systems,** Non-Perimeter
1		Credit 7.1	**Thermal Comfort,** Comply with ASHRAE 55-1992
1		Credit 7.2	**Thermal Comfort,** Permanent Monitoring System
1		Credit 8.1	**Daylight & Views,** Daylight 75% of Spaces
1		Credit 8.2	**Daylight & Views,** Views for 90% of Spaces

Section total: Yes 10, No 5

Innovation & Design Process

Yes	No		Credit
1		Credit 1.1	**Innovation in Design:** 95% construction waste
1		Credit 1.2	**Innovation in Design:** 40% Water Savings
1		Credit 1.3	**Innovation in Design:** 50% Stormwater Capture
1		Credit 1.4	**Innovation in Design:** Exceed MRc4
1		Credit 2	**LEED™ Accredited Professional**

Section total: Yes 5, No 0

brightworks sustainability advisors · portland + san francisco + los angeles · www.brightworks.net

Figure 2.4 Oregon Health and Science University's Center for Health and Healing completed its LEED Platinum certification with a total of 55 points (Platinum requires at least 52), including all 10 Energy Efficiency points and all 5 "Water Efficiency" points. *Courtesy of Brightworks.*

water use, waste generation, etc.) and its basic ingredients (materials and systems). Owners of commercial and institutional buildings have far less knowledge of what is in the building they just built or bought than you might think, because the construction process is pretty messy; there are usually thousands of design decisions made, along with many product and materials substitutions and changes during construction, and there is seldom money left over to document what really went into the building, so the construction documents often give an incomplete or even inaccurate picture of what's actually there and how all of the building systems are supposed to work together.

To understand a building's ingredients and its expected performance (including operating costs for energy and water), an "eco-label" such as the LEED rating is especially valuable both to building owners and to occupants who may naturally be more concerned about how healthy the building is, rather than how much water it saves.

Complicating this rather straightforward percentage method (for determining levels of LEED certification) is the addition of a sixth category with up to five "bonus" points for "innovation and design process" (see Table 2.2). In addition to securing a certain number of points, each rating system has "prerequisites" that each project must meet, no matter what level of attainment it achieves. For example, a LEED-NC–certified building must reduce energy use at least 14 percent below a comparable building that just meets the ASHRAE 90.1-2004 standard (or 10 percent below the newer ASHRAE 90.1-2007 standard).

Table 2.5 shows the four major systems that account for the vast majority of LEED registered and certified projects as of early 2008, not including the LEED for Homes and LEED for Neighborhood Development pilot programs. From this table, you can see that the LEED-CS system is the second-most popular, followed by LEED-EB. For the purposes of this book, we're only going to focus on LEED-NC and LEED-CS, which represent about 80 percent of all LEED registered and certified projects to date.

To best understand LEED, it helps to think of it as a self-assessed, third-party verified rating system. In the case of a LEED certification, a project team estimates the particular credits for which a project qualifies and submits its documentation to the USGBC, which assigns the review to an independent third party. The reviewer has three choices with each point:

1 Agree with you and award the point claimed.
2 Disagree and disallow the point.
3 Ask for further information or clarification.

To resolve differences of opinion, there is a one-step appeal process.

LEED FOR NEW CONSTRUCTION

The most widely known and used LEED system is LEED-NC, which is useful for all new buildings (except core and shell developments), major renovations and housing of four stories and above. Table 2.2 captures the essence of the LEED-NC rating system's major issues. Through the end of 2007, about 68 percent of LEED projects were registered and 74 percent were certified under the LEED-NC assessment method. LEED-NC can also be used for projects on college and corporate campuses and for schools, in which common systems (e.g., parking, transportation, and utilities) often supply a number of buildings.

A LEED-NC rating is typically awarded after a building is completed and occupied, since it requires a final checkout process known as "building commissioning" before the award can be made. Under the current LEED version 2.2, certain credits known as "design phase" credits can be assessed at the end of design and prior to construction, but no final certification is made until all credits are reviewed after substantial completion of the project.

LEED FOR CORE AND SHELL BUILDINGS

LEED for Core and Shell is a system employed typically by speculative developers who control less than 50 percent of a building's tenant improvements at the time on construction. They may build out 40 percent of the space for a lead tenant, for example, and then rent the rest of the building to several tenants who will take much smaller spaces. LEED-CS allows a developer to "pre-certify" a design at a certain level of attainment, then use the LEED rating to attract tenants and, in some cases, financing. Once the building is finished, the developer submits documentation to secure a final LEED rating. Figure 2.5 shows how the LEED credits in the five main categories are distributed in the LEED-CS system. Except for having eight fewer total points (including two fewer points for energy efficiency), the distribution of credits is quite similar to LEED-NC.

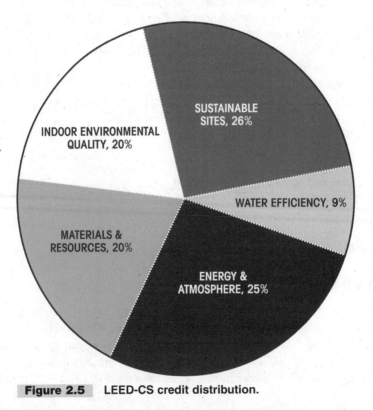

Figure 2.5   LEED-CS credit distribution.

The benefit of the LEED-CS system for a commercial developer is that marketing cannot wait until a building is finished. By allowing a pre-certification using a system very similar to LEED-NC, the LEED-CS rating assists the developer in securing tenants and sometimes financing, and thereby encourages more green buildings. Not only that, LEED-CS awards a point for creating tenant guidelines that encourage each tenant to use the LEED-CI system to build out their interior spaces. If that happens, the result is similar to a LEED for New Construction building, and everyone is happy!

PLATINUM PROJECT PROFILE

Signature Centre, Denver, Colorado

Designed, developed, managed and built by Aardex, LLC, the Signature Centre is a 186,000-square-feet, five-story, class AA speculative office property. The total project cost, including land, finance, and soft costs, was less than $220 per square foot. The Signature Centre was designed to reduce total energy use by a minimum of 36 percent. Indoor water use is estimated to be 40 percent less than a similar conventional building. The building's underfloor air system not only improves indoor air quality and reduces absenteeism, but it also reduces utility costs by 30 percent. The project materials include 20 percent recycled-content materials, 50 percent FSC-certified wood and 20 percent locally sourced materials.*

Courtesy of Aardex, LLC.

---

*John Trojan, "We Want to Raise the Bar," Business Leader Magazine, October 2007, pp 21-27.

There is growing evidence that a LEED-CS certification helps developers to lease space faster and attract better tenants. There is also evidence that ENERGY STAR buildings attract higher rents and result in higher resale values.* In Atlanta, Hines certified their 1180 Peachtree building as LEED-CS Gold.[†] Both the Hines' LEED-Silver One South Dearborn in Chicago and the 1180 Peachtree building were sold in 2006 after completion of construction and leasing activity. Jerry Lea of Hines comments about the benefits of the LEED rating system: "Both buildings got the highest sales price (dollars per square foot) for buildings ever sold in those two markets. Is it because they were green?....I think there is some correlation that green buildings help you lease the space, and that helps sell them."[‡]

LEED FOR COMMERCIAL INTERIORS

LEED-CI is designed mainly for situations in which the base building systems are not changed and which a tenant only takes up a few floors in a much larger building. In this circumstance, the ability to affect total energy and water use, or such issues as open space, landscaping or stormwater management is either much smaller or nonexistent. Thus, other green building measures are incorporated into the evaluation system. These measures include choices that tenants can make about lighting design, energy-using equipment, lighting control systems, submetering, furniture and furnishings, paints, carpet, composite wood products, and length of tenancy.

Because the focus of this book is on the use of integrated design process (IDP) in new building construction, I'm not going to focus much attention on LEED-CI. This is not meant to slight the many fine LEED-CI projects, but the fact is that most tenant improvements happen so fast, there is not time for a conventional IDP.

LEED FOR EXISTING BUILDINGS

LEED-EB was originally proposed and designed to be a method for assuring on-going accountability of LEED-NC buildings over time. It has become instead a stand-alone rating system for building owners who want to benchmark their operations against a nationally recognized standard. LEED-EB addresses many issues not dealt with in new construction, including upgrades, operations and maintenance practices, environmentally preferable purchasing policies, waste management programs, green housekeeping, continuous monitoring of energy use, retrofitting water fixtures to cut use, relamping, and a host of other measures. By early 2008, seven projects had received LEED-EB Platinum ratings, and the system appeared to be gaining momentum, as new LEED-EB project registrations in 2007 almost tripled the number of such undertakings underway in the United States. Again, similar to LEED-CI, we're not going to

---

*See "Does Green Pay Off?" by Professor Norman Miller, www.green-technology.org/green_technology_magazine/norm_miller.htm, accessed July 31, 2008.

[†]www.hines.com/property/detail.aspx?id=507, accessed March 20, 2007.

[‡]Jerry Lea, Hines, Interview, March 2006.

Figure 2.6   **The William J. Clinton Presidential Library and Museum in Little Rock, Arkansas received a LEED-NC Silver certification and later, after some additional improvements, earned a LEED-EB Platinum certification.** ©*Photography by Timothy Hursley.*

devote any time in this book to LEED-EB, because it simply doesn't require an IDP in the same way as new buildings. Figure 2.6 shows the Clinton Presidential Library and Museum originally certified under LEED-NC, then re-certified under LEED-EB.

Other Green Building Rating Systems

In addition to LEED, there are other commercial and institutional green building rating systems. One system in the United States is called Green Globes, a program of the Green Building Initiative. The Green Globes rating system is supposedly easier for project teams to use, but currently has less than 1 percent of the market for commercial and institutional buildings.* However, Green Globes has its adherents, mostly because

*Green Building Initiative [online] www.thegbi.org. As of April 2007, GBI reported only eight projects certified under the Green Globes standard, vs. more than 600 certified by the LEED for New Construction standard at the same time, http://www.oregonlive.com/oregonian/stories/index.ssf?/base/business/1176436580204130.xml&coll=7&thispage=2, accessed July 31, 2008.

it is said that the certification costs are less expensive than LEED's costs. (This claim has not been independently verified.) Because the system is a self-assessment, without a strong third-party review, critics contend that it lacks the rigor and, therefore, the credibility of LEED.

Along with the USGBC, the Green Building Initiative is an accredited U.S. standards development organization. A 2006 study by the University of Minnesota compared the credits offered by the two systems and found 80 percent of the available points in Green Globes are addressed in the LEED-NC version 2.2 (the current standard) and that 85 percent of the points in LEED-NC version 2.2 are addressed in Green Globes.* In essence, the standards are virtually identical, but LEED has market dominance and will likely keep it in the years ahead.

Three non-U.S. rating systems have substantial support in their respective markets: the Japanese CASBEE system, the British BREEAM, and the Australian Green Star.† The main standard used in the United Kingdom, BREEAM is supported by the nonprofit Building Research Establishment and has the longest track record of any rating system. Through early 2008, BREEAM had certified more than 1200 commercial and institutional buildings, about the same as LEED (but in a country one-fifth the size of the United States). However, only LEED is supported by the U.S. federal government. The GSA report mentioned earlier in this chapter compared LEED with Green Globes and these three other systems for rating the "green-ness" of a building design and construction project. Although the study found each of the rating systems has merits, GSA concluded that LEED "continues to be the most appropriate and credible sustainable building rating system available for evaluation of GSA projects."‡

TYPICAL GREEN BUILDING MEASURES

While there's no such thing as a "typical" green building, there are specific design and construction measures that are used in many high-performance buildings. If you are a designer, understanding these measures will help you work with green builders, building owners, developers, facility managers, government officials, business clients, nonprofit executives, or just interested stakeholders in a green building program.

Based on an analysis of the first 1015 LEED for New Construction–certified projects, the following technical measures are those that one might associate with a typical green building project. To illustrate the use of these measures, I prepared an analysis of 25 LEED-NC Platinum and 105 LEED-NC Gold projects (Table 2.6).§ There are more Platinum and Gold projects, of course, but the scorecards are not readily available.

---

*"Green Buildings and the Bottom Line," *Building Design & Construction* magazine, supplement, pp. 56-57, November 2006, available at www.bdcnetwork.com, accessed July 31, 2008.

†U.S. Green Building Council [online], accessed April 22, 2007, https://www.usgbc.org/ShowFile.aspx? DocumentID=1916.

‡U.S. Green Building Council [online], accessed April 22, 2007, https://www.usgbc.org/ShowFile.aspx? DocumentID=1916,accessed April 3, 2007.

§Research by Beth M. Duckles, a PhD candidate at the University of Arizona, based on the USGBC website published data through the end of March 2008, www.usgbc.org, accessed July 31, 2008.

TABLE 2.6 USE OF LEED CREDITS IN GOLD AND PLATINUM PROJECTS

LEED CREDIT CATEGORY	PERCENTAGE OF PLATINUM PROJECTS AWARDED THIS CREDIT (N = 25)	PERCENTAGE OF GOLD PROJECTS AWARDED THIS CREDIT (N = 105)
1. Site restoration (SSc5.2)	88	61
2. Stormwater control (SSc6.1)	96	61
3. Urban heat island effect (Green roof or ENERGY STAR roof, SSc7.2)	84	66
4. Light pollution reduction (SSc8)	76	44
5. Renewable electricity (12.5%/15% of total, EAc2.3)	44	8
6. Measurement and verification (EAc5)	64	42
7. Purchased green power (EAc6)	84	55
8. Recycled-content materials @ 20% (MRc4.2)	84	90
9. Bio-based materials (MRc6)	28	8
10. Certified Wood @ 50% (MRc7)	48	42
11. Carbon dioxide monitors (EQc1)	96	69
12. High-efficiency ventilation @ 30% (EQc2)	68	36
13. Improved air quality at occupancy (EQc3.2)	84	59
14. Underfloor air systems for interior thermal comfort (EQc6.2)	60	37
15. Daylighting (EQc8.1)	84	51
16. Views to outdoors (EQc8.2)	88	68

- Solar photovoltaic systems systems (44 percent of Platinum projects had at least 12.5 percent of total energy use supplied by PV systems, vs. only 8 percent of Gold projects)
- Site restoration (used in 88 percent of Platinum projects and 61 percent of Gold projects, but only 56 percent of Silver projects)
- Carbon dioxide monitors (used in 96 percent of Platinum projects, 69 percent of Gold projects, and 59 percent of Silver projects)

- Green or LEED-compliant roofs (used in 84 percent of Platinum projects and 66 percent of Gold projects, versus 63 percent in LEED Silver projects)*
- Use of certified wood products (in 48 percent of Platinum projects and 42 percent of Gold projects, versus about 19 percent in all LEED Silver–certified projects)†
- Rapidly renewable materials such as cork and bamboo flooring (used in 28 percent of Platinum projects versus less than 5 percent of other projects).
- Daylighting design (used in 84 percent of Platinum projects, 51 percent of Gold projects and only 41 percent of Silver projects).

Many of these systems and approaches aren't common because they have fewer opportunities (e.g., hard-to-restore sites in dense urban areas), experience supply-chain difficulties or require greater initial cost. The primary reason of course for the lack of use of any green building measure is the higher initial cost, followed by the relative inexperience of design teams working with various systems and products.

THE CASE FOR HIGH-PERFORMANCE BUILDINGS

Owners and developers of commercial and institutional buildings across North America are discovering that it's often possible to have "champagne on a beer budget" by building high-performance buildings on conventional budgets. The Harvard Blackstone Renovation was reportedly built to LEED Platinum standards for no initial capital cost increase. The OHSU building in Portland, Oregon, was built for a net cost premium of about one percent, according to the developer. As we'll see in this book, many private developers, building owners and facility managers are advancing the state of the art of sustainable design in commercial and institutional buildings through new tools, techniques, and creative use of financial and regulatory incentives.

Leland Cott of Bruner/Cott & Associates was the lead architect for Harvard's Blackstone Office Renovation, which cost about $10.5 million. Originally the project started out with a Gold goal, one that was contained in the Request for Proposal for architectural services. The goal for a high-performance outcome set in motion a whole range of options, according to Cott, that resulted in the Platinum certification. For example, a further exploration of stormwater management options resulted in a plan to recapture stormwater and add extra landscaping to help with infiltration on site. One of the challenges for upgrading a 100-year-old brick building was how to insulate it, to improve comfort and energy savings. The team ended up putting icynene insulation on the inside so that moisture could continue to move in and out of the building according to the season, without condensation on interior walls. According to Cott, Harvard reports 40 percent energy savings in the renovated building. The project included a strong charrette that developed a mission statement and a "green pledge."

Cott believes that the advent of Building Information Modeling (BIM) software will be a very progressive tool for achieving sustainable outcomes, since it will be easier

*Jerry Yudelson, *Marketing Green Building Services: Strategies for Success*, 2006, (Amsterdam: Elsevier/ Architectural Press), p. 129.
†Ibid.

to do early stage energy modeling, for example, and then to change things without incurring large additional design costs. BIM also helps the architect as lead designer to spot conflicts earlier, for example, between mechanical and structural systems, and resolve them before too many costs are incurred. One reason these conflicts can occur is because usually a building's architectural and structural systems are defined well in advance of its mechanical and electrical systems, with only standard "rule of thumb" space allowances reserved for the latter. This makes it harder to innovate with mechanical systems, such as chilled beams, that require different space allowances.*

PLATINUM PROJECT PROFILE

Blackstone Office Renovation, Harvard University, Cambridge, Massachusetts

Three nineteenth century masonry buildings were renovated to create a workplace for Harvard University's Operations Services group. The 40,000-square-feet Blackstone Station was completed in May 2006 without any increase in initial

Photography by Richard Mendelkorn.

---

*Autodesk White Paper: http://images.autodesk.com/adsk/files/whitepaper_revit_systems_bim_for_mep_engineering.pdf, accessed April 30, 2008.

construction costs (compared to a traditional building renovation). Blackstone reduces energy use by 42 percent compared with a code building, in part due to the daylight sensors, occupancy sensors and an energy-efficient elevator. Two geothermal wells provide cooling. Dual-flush toilets, water-free urinals, and low-flow sinks and showerheads reduce occupant water use by 43 percent (compared to code). A bioretention pond and bioswale reduce and treat water runoff. Over 99 percent of the construction waste was reused or recycled.*

TO LEED OR LEAD?

LEED has gained prominence as the preferred certification system for larger projects (although both may be used by developers and building owners) because it focuses on a broader range of issues than most energy efficiency guidelines. For example, if owners' points of focus are primarily on energy use, reducing carbon dioxide emissions (linked to global warming) and improving indoor air quality, a variety of guidelines such as the proposed ASHRAE Standard 189P can take them there efficiently.[†] These improvements lead to reducing operating costs and improved occupant health, productivity, and comfort. *However, at this time, only LEED and ENERGY STAR have marketplace acceptance at this point as "brand names" that indicate a high level of performance against measurable criteria.*

Both LEED and other building evaluation systems encourage an integrated design process, in which the building designers (mechanical, electrical, civil/structural, and lighting engineers) are brought into the design process with the architectural and interiors teams at an early stage, often during programming and conceptual design. Integrated design explores, for example, building orientation, massing and materials choices as critical issues affecting energy use and indoor air quality, and attempts to influence these decisions before the basic architectural design is fully developed.

DESIGNING HIGH-PERFORMANCE BUILDINGS

What are the design and operating characteristics of today's high-performance buildings? They save 25 to 50 percent (or more) of conventional building energy use by incorporating high-efficiency systems and conservation measures in the basic building envelope, HVAC plant, and lighting systems. These systems and efficiency measures can include extra insulation, high-quality glazing, and solar control measures; ENERGY STAR-rated appliances such as copiers, computer

---

*"Harvard Green Campus Initiative: High Performance Building Resource," May 2006, pp 1-8. Consigli Construction Co., Inc. "Web Exclusive: 19th Century Platinum," Environmental Design & Construction, December 3, 2007 [online], http://www.edcmag.com/CDA/Articles/Web_Exclusive/BNP_GUID_9-5-2006_A_1000000000000214697, accessed April 2008. Bruner/Cott [online], http://www.brunercott.com/ library/hublackstone/blackstone.htm, accessed April 2008.

[†]Standard 189P "will provide minimum requirements for the design of sustainable buildings." See www.ashrae.org/pressroom/detail/13571, accessed July 31, 2008.

monitors, and printers; building orientation and massing to utilize passive solar heating and cooling design; high-efficiency lighting (often using high-output T-5 lamps in many applications); carbon dioxide monitors that monitor room occupancy and adjust ventilation accordingly, so that energy is not wasted in ventilating unoccupied space; occupancy sensors—which turn off lights and equipment when rooms are unoccupied; and higher-efficiency HVAC systems, variable speed fans and motor drives, to produce the same comfort level with less input energy; and many similar techniques.

The New Buildings Institute conducted a study in 2007 of more than 120 LEED-certified office buildings for which energy performance data were readily available.[*] LEED Gold and Platinum projects had an average energy use about 44 percent below the average of all commercial buildings in a 2003 national survey. Overall, for all LEED projects, the energy savings amounted to 24 percent below the average energy use of commercial buildings. The study concluded, "On average, LEED buildings are delivering anticipated savings," but the data showed a large amount of variation, "suggesting opportunities for improved programs and procedures."

Before the end of construction, LEED requires that all buildings to be commissioned, through the use of performance testing and verification for all key energy-using and water-using systems. Typically, commissioning involves creating a plan for all systems to be tested, performing functional testing while the mechanical and controls contractors are still on the job, and providing the owner with a written report on the performance of all key systems and components. Green building commissioning involves third-party peer reviews during design, to see if design intent has actually been realized in the detailed construction documents. Finally, most commissioning programs also involve operator training and documentation of that training for future operators. Getting the future building maintenance staff involved is also a critical component of effective commissioning practice.[†]

Think of commissioning as analogous to the "sea trials" a ship undergoes before it is handed over to the eventual owners. No ship would be put into use without such trials, which may expose flaws in design or construction. In the same way, no building should begin operations without a full "shakedown cruise" of all systems that use energy and affect comfort, health, and productivity. Often, the documentation provided by the commissioning process can be helpful later on in troubleshooting problems with building operations. It's really amazing to me that any building would be built today without a full commissioning process, so it's a good thing, absolutely essential for a high-performance building, that LEED requires it for all projects.

High-performance buildings achieve higher levels of indoor air quality through a careful choice of less-toxic (low-VOC or no-VOC) paints, sealants, adhesives, carpets, and coatings for the base building and tenant improvements, often in conjunction with building systems that provide higher levels of filtration and carbon dioxide monitors

[*]Cathy Turner and Mark Frankel , "Energy Performance of LEED for New Construction Buildings," New Buildings Institute, March 2008, www.newbuildings.org, accessed April 30, 2008.
[†]Personal communication, Paul Schwer, PAE Consulting Engineers, May 2008.

to regulate ventilation according to occupancy. With so many building occupants today having breathing problems and chemical sensitivities, it just makes good business sense to provide a healthy building. Documentation of these measures can often help provide extra backup when fighting claims of "sick building syndrome." This benefit of "risk management" is an often overlooked aspect of green building guidelines, but can often be useful to demonstrate to prospective tenants or occupants the often "invisible" measures taken by building designers and contractors to provide a safe and healthy indoor environment.

Healthy buildings incorporate daylighting and views to the outdoors not only for occupant comfort, health, and productivity gains (Fig. 2.7), but also to reduce energy costs. There is a growing body of evidence that daylighting, operable windows, and views to the outdoors can increase productivity from 5 to 15 percent and reduce illness, absenteeism, and employee turnover for many companies.* Throw in higher levels of building controls that allow for such things as carbon dioxide monitoring and demand-controlled ventilation adjustments, for example, and one has an effective program addressing the "people problem" that can be sold to prospective tenants and

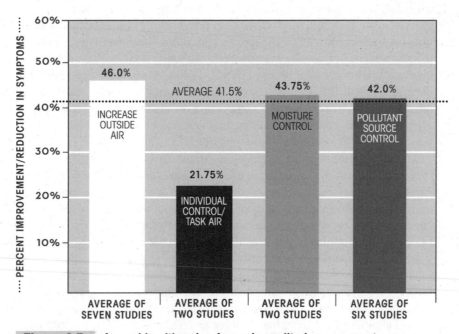

Figure 2.7   **Annual health gains from air quality improvements.**

Center for Building Performance and Diagnostics, Carnegie Mellon University. BIDS™: Building Investment Decision Support Tool

*See for example, studies by the Heschong Mahone Group for Pacific Gas & Electric Company and the California Energy Commission, available at www.h-m-g.com, accessed July 31, 2008.

other stakeholders. For owner-occupied buildings, these savings alone are often enough to justify the extra costs of such projects. Considering that 70 percent or more of the operating costs of service companies (which most are) relate to employee salaries and benefits, it just makes good business sense to pay attention to productivity, comfort, and health in building design and operations.

Consider what Ben Weeks of Denver's Aardex says on this subject about the benefits of their Signature Centre office project.*

In one specific case, in Albuquerque, New Mexico, we built an office space, and the Social Security Administration's Office of Hearings and Appeals occupies about 28,000 feet of a facility designed for their specific use. They had come out a previous space where they had been disjointedly housed in several different locations within the same building. From before and after analysis of their case management, we identified that by going from that location to the new facility their productivity increase was 78 percent, an enormous increase. That is almost incredible. When you think that business devotes between 80 and 90 percent of its total operating expense to people and roughly 5 percent to office rent, if the space can influence the productivity of the tenants—even just a little, it makes an enormous difference.

Typically, businesses will pay anywhere from $300 to $600 per square foot for its people. And they'll pay $15 to $20 per square foot for the office space. If a building can increase the productivity of the people by even 5 percent, that's significantly more than the total cost of the office rent. We realized this effect and that's why we developed the name "User Effective Buildings" for our approach to real estate development. It describes the design relationship between a building design, the people that occupy the building and the effect on productivity.

We have done a lot research into the aspects of buildings like daylight harvesting (Fig. 2.8), internal artificial light management, environmental controls, indoor air quality controls and those sorts of things, considering the impact that those design elements have on the human experience and thereby productivity of the occupants. We've incorporated optimal outcome thresholds into our guiding principles for our design strategies. When that is done, we believe we can develop a building in such a way that it will house the current and future occupants effectively far into the future. They should be able to maximize their useful and productive enjoyment of those buildings. We believe that the numbers, the financials, the mechanics of it, are such that the information and the data are irrefutable. Although the example of the Social Security Office of Hearings and Appeals and their 78 percent increase in productivity is one that we will anecdotally discuss when talking to people, we don't ever suggest that any company might have the potential to increase by 78 percent because leaders don't like to think that their people are currently working at 22 percent or less of capacity.

*Interview with Ben Weeks, Aardex, March 2008.

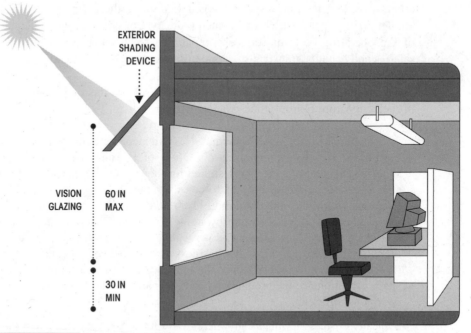

EXTERIOR
SHADING
DEVICE

VISION
GLAZING

60 IN
MAX

30 IN
MIN

Figure 2.8   **Good daylighting design provides natural light for offices without glare and unwanted heat gain.**

Looking to the Future

The U.S. Environmental Protection Agency's ENERGY STAR program, the most well-known to consumers, should also be used to promote energy-efficient and Zero-Net-Energy, or *carbon-neutral,* commercial and institutional buildings. By 2010, we will begin seeing buildings routinely designed to cut energy use 50 percent or more below 2005 levels through integrated design and innovative technological approaches. With the growing awareness of the carbon dioxide/global warming problem and the contribution of buildings and urban settlement patterns to this observed global warming, architects and others in the design and construction industry have begun to propose positive actions. One sign of this is the position statement adopted by the American Institute of Architects (AIA) in December 2005, calling for a minimum 50 percent reduction in building energy consumption by 2010.* In its statement, the AIA supported "the development and use of rating systems and standards that promote the design and construction" of more resource-efficient communities. This position statement echoes the requirements of the "Architecture 2030 Challenge," which seeks to reduce building energy use by 90 percent by 2010.[†]

---

*American Institute of Architects [online], December 19, 2005 press release, viewable at www.aia.org.
†See www.architecture2030.org for regular updates on this challenge, accessed July 31, 2008.

Many cities have subscribed to climate change initiatives and will begin to require green buildings for residential projects, especially large developments with major infrastructure impacts. For example, by early 2008, more than 800 mayors representing cities in all 50 states and Washington, D.C. signed on to a climate change initiative.* Mayors who sign on to the agreement make a commitment to reduce greenhouse gas emissions in their own cities and communities to a level 7 percent below 1990 levels by 2012 through a series of actions like increasing energy efficiency, reducing vehicle miles traveled, maintaining healthy urban forests, reducing sprawl and promoting use of clean, renewable energy resources. In 2008, both Los Angeles and San Francisco passed ordinances requiring private sector projects above 50,000 square feet to achieve LEED certification.

Beginning in 2004, many states, large universities, and cities began to require LEED Silver level (or better) achievements from their own construction projects. Many universities have instituted LEED Gold requirements for large capital projects, among them Arizona State University (ASU).[†] Michael McLeod is Facilities Director for ASU's new Biodesign Institute, the first two buildings (of four planned) of which have been awarded LEED Gold and Platinum certifications. Of the approach to LEED in these projects, McLeod says[‡]:

> The interesting part was that not everyone bought into the whole LEED program at first. There were people who were skeptical of it within the university. Luckily, we had already selected Gould Evans and Lord Aeck Sargent as our designers. They were both very interested and motivated for LEED and sustainability. To be honest, I found that a lot of consultants still haven't gotten on the bandwagon. The contractors, believe it or not, seem to be jumping on it left and right. I've been a little less than impressed with a lot of the consultants.
>
> The team was not selected based on their experience in sustainability. When they were all selected, LEED certification wasn't even in the picture or it was a very weak possibility. The take away from that is if you hire good, high-quality contractors and consultants, then they can rise to the occasion and do what you want. We were either really lucky or really good in the selection because both the contractor and the consultants really jumped on it. They worked hard to make it a success.
>
> These buildings were part of a bigger study that was done a few years ago to develop research at ASU—to motivate it, increase it and move us a few tiers higher in the research community. The fundamentals of the building were set at that point. When we brought on the design architects, we did not really do a charrette per se. We started

---

*City of Seattle, Mayor's Office [online], accessed April 26, 2007, http://www.seattle.gov/mayor/climate/PDF/USCM_Faq_1-18-07.pdf.

[†]Association for the Advancement of Sustainability in Higher Education (AASHE) [online], www.aashe.org/resources/pdf/AASHEdigest2006.pdf, accessed April 26, 2007. The AASHE website is a great place to keep track of the push for sustainable buildings in higher education, especially the American College and University President's Climate Committee, signed by more than 500 presidents, www.presidentsclimatecommitment.org.

[‡]Interview with Michael McLeod, March 2008.

building this project without knowing who was going to be in it. In fact, the director, Dr. George Post, had not even been hired yet. He came in shortly thereafter and was very quick to come up to speed and give us his vision but fundamentally when we started, we didn't know who was going to occupy the building.

When that's the case, you build a basic, flexible facility, which is what they wanted in the first place. Then you're not influenced by a lot of individual researchers who want it the way they always had it. It allowed us to break out of box. We set up the labs and did what I call the minimums. That way there was enough infrastructure in the lab so when the researchers came in we retrofitted it to their specific needs. It was set up to handle everything from optics and lasers to chemists, physicists, and engineers—and we've got them all. We can set up the lab areas to accommodate them.

After the first building was built and occupied, the whole LEED movement got going. We decided because ASU is the premier institution in the state, that all university buildings should be LEED certified. The administration gave me the directive to go back and certify Building A. Luckily, we had an excellent design and we had done some of the big dollar things already, for example, commissioning. In fact, we commission all of our buildings. A lot of people include commissioning as a cost of LEED, but we do it because we believe it's the right thing to do. I went back and filled out the LEED application and looked at some modifications that were needed—we had to add carbon dioxide sensors and some other things. It cost just a little over $250,000. Basically, it was just a good, solid, design and a well-built building. I would say 85 percent of the exercise was filling out the application and going through the process. It shows that if you do a good building, getting a LEED certification is not as a big of a deal that a lot of people make it out to be.

PLATINUM PROJECT PROFILE

Biodesign Institute, Phases 1 and 2, Arizona State University, Tempe, Arizona

The Arizona Biodesign Institute consists of two LEED-certified buildings—one Gold (Building A) and the other Platinum (Building B). Connected by glass walkways, together the two buildings contain 350,000 square feet of offices, open labs, plus dining, and auditorium facilities on three floors. The construction cost of the Institute was $104 million and the total project cost was $160 million. Remote sensors can detect air pollutants in a particular zone and cue the system to flush out the air as needed, minimizing energy consumption (by reducing the number of ventilation air changes) while meeting the strict fresh-air requirements for lab buildings. Rainwater and air-conditioning condensate are collected in a 5000-gallon cistern for reuse in landscaping. An internal louver system automatically tracks the sun's path, diffusing sunlight by reflecting it off the interior ceilings. A 167-kilowatt photovoltaic array contributes to

Courtesy of The Biodesign Institute.

building B's total energy reduction of 58 percent. Fifteen percent of the building's materials contain recycled content.*

The Larger Picture

Reducing carbon dioxide emissions from the buildings sector is critical to our ability to combat global warming. Energy efficient design and operations of buildings, along with onsite renewable energy production are a strong part of the answer to the challenge to

*Green Giant: Biodesign Institute Goes Platinum, Arizona State University, July 31, 2007 [online], http://asunews.asu.edu/node/689, accessed April 2008. David Sokol, Arizona State University Biodesign Institute, Green Source [online], http://greensource.construction.com/projects/0801_ArizonaStateUniversity.asp, accessed April 2008.

Americans to reduce their "ecological footprint."* Green buildings are an important component in the effort to bring carbon dioxide emissions back to 1990 levels, as required by the Kyoto Protocol, so that we can begin to stabilize carbon dioxide concentrations in the atmosphere at levels no more than 20 percent above today's. Recent studies by the international consulting firm McKinsey indicate that buildings can provide up to 25 percent of the required carbon emission reductions and at costs that can be easily recovered over the life of the building.[†]

Barriers to Green Building Growth

There remain barriers to the widespread adoption of green building techniques, technologies, and systems, some of them related to real-life experience and the rest to a lingering perception in the building industry that green buildings add extra costs. Senior executives representing architectural and engineering firms, consultants, developers, building owners, corporate owner-occupants, and educational institutions have positive attitudes about the benefits and costs of green construction, according to the 2005 *Green Building Market Barometer*, a survey conducted by Turner Construction Company.[‡] For example, 57 percent of the 665 executives surveyed said their companies are involved with green buildings; 83 percent said their green building workload had increased since 2002; and 87 percent said they expected green building activity to continue. However, despite an overwhelming sense that green buildings provided considerable benefits, these same executives thought that green buildings cost 13 to 18 percent more than standard buildings!

In a 2007 survey by *Building Design & Construction* magazine, 41 percent of construction industry participants surveyed said that green buildings added 10 percent or more to cost of buildings, even while the clear evidence is that cost increases are less than 10 percent![§] I address the cost issue in Chap. 7 but one should never forget that all building decisions are fundamentally economic ones and that the added costs of high-performance buildings will continue to be an issue, until project teams figure out how to design such projects consistently with no cost increase.

---

*See www.footprintnetwork.org, for a fuller explanation of the term, ecological footprint, accessed July 31, 2008.
[†]"A Cost Curve for Greenhouse Gas Reduction," McKinsey Quarterly, February 2007, www.mckinseyquarterly.com/Energy_Resources_Materials/A_cost_curve_for_greenhouse_gas_reduction_1911, accessed June 30, 2008.
[‡]Turner Construction Company [online], www.turnerconstruction.com/greensurvey 05.pdf, accessed March 6, 2007.
[§]Building Design & Construction magazine, 2007 Green Building White Paper, November 2007, at page 8. Available at www.bdcmag.com, accessed July 31, 2008.

3

THE PRACTICE OF

INTEGRATED DESIGN

So far, I've presented the case for green buildings and shown some outstanding examples of high-performance building design. Let's turn now to the subject of how design teams are actually achieving high-performance results. The basic thesis of this book is that *systems are more powerful than individuals*. If design talent is unevenly distributed, as it surely is, then we need to redesign our project delivery process to get better results. We can't rely on superior design talent alone to get superior results. That's why the integrated design process is so important. Otherwise we're back in the world of the movie *Groundhog Day*, where nothing ever changes. The push for LEED Gold and Platinum high-performance buildings is the driving force, along with the growing emphasis on carbon-neutral solutions, that gives us a "once in a lifetime" opportunity to change our project delivery approach for the better. As Leith Sharp wrote in the Foreword, effective integrated design can produce significant innovations and cost savings simultaneously.

Elements of the Integrated Design Process

The key elements of the integrated design process are simple and straightforward. What's not obvious is how to implement the process in actual practice. For a high-performance building project, the process consists of the following steps:

1 Make a commitment to integrated design and hire design team members who want to participate in a new way of doing things. From the owner's perspective, this may mean having to accept new consultants who may not be as familiar with a campus, development, or institutional setting. Your current "favorite" consultants may not be willing to commit to the process.

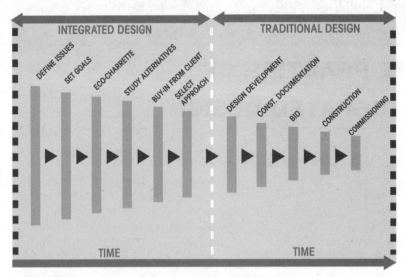

Figure 3.1 The opportunities for integrated design team diminish over time, so the process must be front-end-loaded.

2 Set "stretch" goals for the entire team, such as LEED Platinum, Living Building certification (see Chap. 14) or Zero Net Energy, and judge the final result from that standpoint.

3 Get the team to commit to zero cost increase over a standard budget, so that cost management is a consideration from the beginning and the need to find "cost transfers" or "cost tradeoffs" is built into everyone's thinking.

4 "Front load" the design process with environmental charrettes, studies and similar "thinking" time (Fig. 3.1). This gets more difficult if the schedule is compressed, but is essential for the process to work.

5 Allow enough time for feedback and revisions before committing to a final design concept. This means that the client has to accept either slightly higher design fees that includes early stage studies or has to accept that more money will be spent during the conceptual and schematic design stages.

6 Everyone has to buy in and participate. No building team member should be allowed to consider just their own special interest. This might mean that the electrical engineer who's going to be responsible for lighting design also has to be concerned with the paint colors (that affect internal reflected light), the glazing (for daylighting) and similar issues.

An Architect's Perspective

Bill Reed writes about the importance of using charrettes to get people on the building team to change practice habits. In his view, the purpose of "integrative design" is to get people to change entrenched patterns that inhibit creativity, systemic thinking, and innovative designs.

The most successful process we've employed to help people to change is this: at the first goal-setting charrette, we map out a design process that shows how people are going to be integrating and communicating, when they're going to be talking and why. It's a very detailed map. It's not a "critical path;" it's an integration roadmap. Without that roadmap, people will fall back into usual practice patterns.*

According to Reed, the basic elements of integrated design include the following activities:

- The client's primary, financial decision maker is in the design process.
- The client selects a design team with the right attitude, one of being willing to learn. (Reed has told me that difficult situations can often occur on projects where the client has already hired a "starchitect," a famous designer who will not commit to participate in a team process or even to attend all the key early project meetings; in that case, as a consultant, Reed sometimes refuses to work on such a project, because the odds for a successful integrated design outcome are much smaller.)
- Stakeholders and the project team spend time aligning expectations and purposes.
- Specific goals are set for a range of environmental targets, even within a LEED certification goal (e.g., we're going to save 50 percent of the energy of a conventional building, regardless of the LEED rating).
- The client and the design leader identify process champions to uphold these goals throughout the design and construction effort.
- System designs are optimized early in the process, using an iterative process in predesign and schematic design phases.
- The design team commits to follow through all the way to the end of construction.
- The project is commissioned to make sure that all systems perform as designed.
- There is ongoing monitoring and maintenance, to ensure that the project attains its desired performance goals.[†]

For Reed, integrated design is all about changing mental models, dominant patterns of seeing the world and paradigms for performing in certain ways. He believes it's critical to break up the narrow boundaries of specialists to return to a more holistic way of viewing design. According to Reed, this is very difficult: "shifting the nature and practice of design from a linear, simplistic cause-and-effect process to one that considers issues from multiple and interrelated systems perspectives is resisted more than any other aspect of green design."[‡]

One thing Reed emphasizes is that the process must be "front-end loaded" with multiple iterations of design ideas and thorough explorations of possibilities happening quickly in the first 2 months of a project. Figure 3.2 shows how the potential cost-effective opportunities for integrated design diminish rapidly over the course of a project.

---

*Interview with Bill Reed, February 2008.
[†]Bill Reed, "Integrated Design," May 8, 2005, private memorandum.
[‡]Bill Reed, "The Integrative Design Process—Changing Our Mental Model," April 20, 2006, private memorandum.

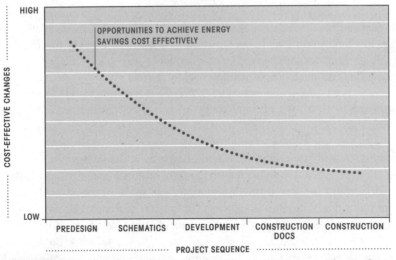

Figure 3.2   Opportunities for cost savings decrease as the project goes along, while the cost for making changes increases dramatically. This is sometimes called the McLeamy Curve, after architect Patrick McLeamy.

What Integrated Design Is Not

Sometimes, it's helpful in understanding a slippery concept such as integrated design by describing what it's NOT. A leading academic text on design process describes what integrated design is not, in this way.*

Integrated design is not necessarily "high tech" or specialized technical design, although it may incorporate such elements, especially for larger projects. The focus instead is on the long-term functioning and health of an entire building system or group of systems, not just specific elements, so the interrelation of the various elements is key.

Integrated design is not some sort of a traditional "hand off" or sequence of activities, proceeding linearly from owner to architect to engineer to general contractor to subcontractor to occupant; instead there are built-in feedback loops as each step of design is evaluated against the project's goals. (This is a point I make many times in this book.)

Integrated design is not simply design by a committee of peers. Recall the adage that "a camel is a horse designed by a committee." There is still a requirement for a design leader; however, that leader needs to genuinely welcome design input from all team members.

---

*Adapted from David Posada, in Alison G. Kwok and Walter T. Grondzik, *The Green Studio Handbook*, 2007, Amsterdam: Elsevier/Architectural Press, pp. 16–17.

It is not another "buzz word" that still follows a more or less conventional design process, with perhaps an eco-charrette thrown in to give a semblance of design integration; it requires instead rethinking of all relationships and purposes, in the interests of a greater goal: sustainable design, construction, and operations practices.

It is not about "chasing LEED points." LEED is an evaluation system that is useful in guiding some design decisions, but does not explicitly require integrated design. This situation frequently finds expression in owners questioning why they need to install bicycle racks and showers to get one LEED point. When this comes up, I always point out that the goal (of bike racks and showers) is to give people an option of riding to work instead of driving; as gas prices march toward $8 per gallon, as many are now predicting, one can expect bicycle commuting to increase. I also like to answer such objections that showers are a useful amenity in their own right, since many people like to run or bike during the lunch hour.

Integrated design is not as easy as changing your shirt every day; old habits die hard. To me, it appears that air-conditioning has made mechanical engineers reactive for decades, because no matter how the architect designs the building, they can still provide more or less adequate comfort by adding air-conditioning tonnage. There are also the risks of trying new things; every departure from "normal" design practices, no matter how intelligent, runs the risk of a lawsuit if things don't work out as planned. To make integrated design work, the team often has to challenge prevailing codes. This is how progress is made, but it isn't easy or fast.

PLATINUM PROJECT PROFILE

Ronald McDonald House, Austin, Texas

The 28,500-square-feet, 4-story Ronald McDonald House provides a home-like environment allowing families to stay together while their children receive medical treatment in Austin-area medical centers. Fifty-four photovoltaic panels provide 10.8 kilowatts of electricity which will power about half of the 30 guest rooms. Each guest room has a dedicated fan coil unit which resides in unoccupied mode until activated by a room key, preventing unnecessary energy use during unoccupied periods. The project received an innovation credit for the combined heating and plumbing methodology utilized in the HVAC system which allows the building's documented energy cost savings to increase from 47 percent to just over 65 percent. Developed on a brownfield, the project also houses administrative offices and common areas as well as butterfly and rooftop gardens, a children's playground and a picnic area.*

---

*http://www.rmhc-austin.org/repository/images/LEED Platinum.ppt, http://www.rmhc-austin.org/repository/pdf/ pd RMHC LEED Brochure REV 12-10.pdf, http://austin.bizjournals.com/austin/stories/2008/05/26/daily8.html, accessed June 2, 2008.

The Role of BHAGs

I'm especially fond of the term "BHAG," or Big, Hairy, Audacious Goal, a "technical" term from the world of management consulting that describes one essential activity that must occur to create high-performance buildings: establishing "stretch" goals for the design team. We all know from watching children over time that the more that's demanded, the better the child responds (up to a point, of course). The late-1980s movie *Stand and Deliver* depicted a group of kids in Los Angeles who became national whizzes in calculus because a determined teacher demanded that they "stand and deliver" the absolute best of which they were capable. Why can't we do just as well with highly educated, highly motivated architects and engineers? Is it because we (or the clients) don't challenge them enough? Is it because we accept the mediocrity of the process as the natural order of things?

Portland, Oregon architect Phil Beyl describes how establishing stretch goals with a sophisticated developer client resulted in a large LEED Platinum building coming out of what could have been a very ordinary medical "build to suit" project.*

We established a LEED Platinum benchmark for the [Oregon Health & Science University's] Center for Health and Healing [Fig. 3.3] before we started doing much design work on the project. That was done for a couple of reasons. One, the project was the first in an emerging neighborhood where the city has great aspirations for it to be a model for the rest of the world for sustainability. So the city had specific interests in pushing the envelope there, as did the client [OHSU]. It was their first building for a new market segment. They also wanted to demonstrate to the city, the state, and the country that sustainability was very high on their agenda for this and all of their [future] facilities. So this first project at the South Waterfront area provided a great chance to demonstrate that commitment.

Second, this was a unique building that had never been done before in the world, to our knowledge. Having those two factors setting the benchmark for performance very, very high really forced the entire team to work in a highly integrated fashion from the very early stages of design. The building orientation, for example, had a huge impact on our capacity to control the cooling loads. If we had to wait and find out down the road [of the design process] how different mechanical systems needed to be integrated into the building and where they needed to be placed, we would have never been able to accommodate them.

I suppose that it's hard to say that there's a better project that I could reference that utilized truly an integrated design process from the very early stages. You always need to integrate the design of all of those disciplines in executing a project, but usually you know enough about basic needs that you can go a ways down the road with the architecture of the building before you need to say, "Exactly how big does that fan need to be? Where does it go?" Because you've made some fundamental

---

*Interview with Phil Beyl, GBD Architects, February 2008.

Figure 3.3   When the Oregon Heath & Sciences University's Center for Health & Healing in Portland received LEED-NC certification, it was the largest Platinum-rated project to date at 412,000 square feet. © Uwe Schneider, www.uweschneider.com.

accommodations for it, chances are it's going to work. But, this project was different. This was off the charts from the get-go, so we needed to know things like that way up front.

Consider another example of the importance of engineers to integrated design. Steve Straus is a mechanical engineer and president at Glumac, one of the west coast's leading building engineering firms, with nearly 50 LEED projects either finished or underway, and a frequent collaborator of Beyl's. He says:*

The key to integrated design is finding an owner and architect that value and appreciate the engineer's input early on in the process and then finding an engineer that actively wants to participate.

Once you're engaged on a project, it's important for the engineer to provide information, be proactive and not wait to be asked for it. [In terms of reducing energy use, for example,] a simple energy model is one way. Secondly, developing an energy pie [chart] will help the team understand where the energy in the building going to be used and what can we do to reduce it [Fig. 3.4]. I think also developing conversations between all of the different engineering teams is important

---

*Interview with Steven Straus, Glumac, February 2008.

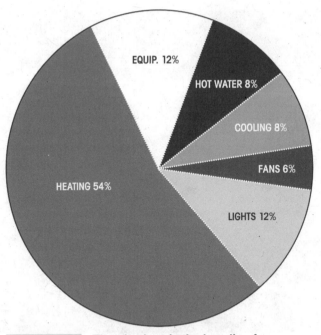

EQUIP. 12%

HOT WATER 8%

COOLING 8%

FANS 6%

LIGHTS 12%

HEATING 54%

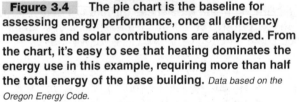

Figure 3.4 **The pie chart is the baseline for assessing energy performance, once all efficiency measures and solar contributions are analyzed. From the chart, it's easy to see that heating dominates the energy use in this example, requiring more than half the total energy of the base building.** *Data based on the Oregon Energy Code.*

because it's not just about the mechanical and electrical systems, it's also the structural engineer and the glazing consultant. It's about getting the communication started so everybody understands how the energy is being used, what the energy pie looks like and what can we all do to contribute to reducing the overall size of the energy pie.

I cannot emphasize too strongly that the design professions have a strong responsibility to work with clients to establish adventurous goals for projects, especially in the energy performance arena, if the industry is ever to have a chance to solve the challenging issues of global warming. Even the Architecture 2030 goal of 50 percent savings by 2010 (compared with 2005 averages) may prove to be too modest.* We really need to begin to design projects with the goal of completely eliminating carbon emissions from energy use, not only at the building site but also at the source, the power plant. About three-quarters of all electricity produced in the United States is used in buildings, so it's in building design that the solution must lie.

*Architecture 2030, www.architecture2030.org.

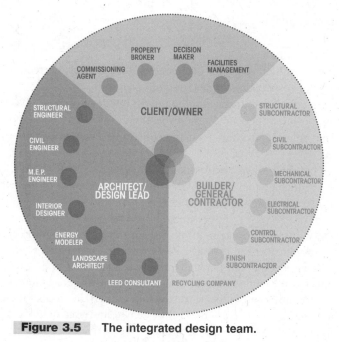

Figure 3.5   **The integrated design team.**

The Integrated Design Team

The four key participants of every integrated design team are the building or project owner (or its designated representative), the architect, the mechanical engineer (including typically HVAC and plumbing) and the builder or general contractor. However, most successful high-performance projects have strong inputs from a variety of other participants, depending on the nature and complexity of the project, the specific sustainable goals sought, and local site and community conditions.

The integrated design team (Fig. 3.5) consists of a wide range of specialties and functions including:

- Architect
- General contractor
- Stakeholders from the owner's side, including project manager
- Structural engineer/façade consultant (for large buildings)
- Mechanical engineer
- Civil engineer
- Electrical engineer/technology consultant
- Landscape architect
- Interior designer

- Lighting designer/consultant
- Energy expert
- Cost management consultant
- Specialized design consultant (e.g., for data centers, laboratories, corrections)
- Mechanical contractor (especially for design/build projects)
- Commissioning authority/agent (especially at the design development phase)
- Natural resources specialist (depending on project)
- Onsite waste management consultant (if using constructed wetlands, for example)

Sally Wilson is global director of environmental strategy for CB Richard Ellis, the world's largest property management firm. She brings the perspective of the commercial real estate broker to the integrated design team.*

Speaking specifically of tenant brokers, a lot of them think the design process is something that the architect or the engineer deals with. It's critical that you lock some of the LEED components into the lease, otherwise it's going to be an uphill battle for the architect or the engineer to get documentation or to get a commitment. The building has to commit to providing infrastructure for certain pieces [of LEED certification]. For example, if you're a multi-floor tenant and you want the water credit and you have to change out the bathroom fixtures to water-free urinals, they're probably not going to let you do that. But if it's the lease, they're going to let you do that. If you negotiate it early as a requirement for the tenants, then you can get that change.

So, here's another participant (the real estate broker) who may need to be integrated into the design process, especially for commercial real estate projects and especially in multi-tenanted buildings such as the 1600 or so commercial development projects now underway seeking certification under the LEED for Core and Shell (LEED-CS) standard.

Integrated Design from the Engineer's Perspective

Geoff McDonnell is a mechanical engineer in British Columbia. Here's how he describes the integrated design process, from an engineer's viewpoint.[†] You can see that the integrated design process can get fairly intricate in a hurry. The real key is that everyone is engaged early as a team, goes off and does their specialized work of analysis and design, then gets back together and tests their ideas against the larger

---

*Sally Wilson, CBRE, Interview, February 2008.
[†]Personal communication, Geoff McDonnell, Omicron, April 18, 2008.

project goals, continually repeating the process until the team as a whole arrives at a final design.

1 The building team should be selected as a group, starting with a blank sheet of paper. If the architect already has a design in mind, it's much harder to do integrated design with everyone contributing as equals. The design team can then assess the site conditions, the local design and building code requirements, and local sources of materials for the building project. The owner's initial maximum budget and building intentions (including project scope) should be established clearly at this stage.

2 The design team and the building owner work together to establish initial project goals and decide on the make-up of the subsequent design charrettes and design team members required for the next steps. At this point, the team creates a statement of the design issues that the project goals would generate and determines the baseline design criteria for the building site, occupancy, and operation and maintenance requirements.

3 The design team assembles the initial building toolbox (using input from each specialty's toolbox, such as radiant cooling systems from the mechanical engineer's), perhaps through one or more "what if" sessions, to see what building design issues should be prioritized to meet the client's goals. At this point the team seeks to resolve major interactions of energy use, occupant comfort, reduced maintenance and life expectancy of building systems. This is a good point to engage the owner's building operations and maintenance group to share their experience with similar projects.

4 Then the team should engage the energy modeler, the cost consultant, the owner's maintenance and operations representative(s), and the construction representative to create simple models of each system. The team works together to design and optimize the right set of systems and building design details that best meet the project goals. The team should expect to generate at least two or three major iterations in building design concepts, with perhaps one or two concepts set aside for further detailed design review. Several options can also be tested at this point, to study HVAC systems types, lighting design and control systems, daylighting options, natural ventilation strategies, and building envelope choices.

5 The design team works with the prospective occupants to establish their requirements and interior spaces, adjacencies, and other programming requirements. The design team then has an opportunity to test the building design toolbox established in the previous step, against what the building occupants expect.

6 In design development phase, the design team members return to their individual disciplines to design a more detailed set of systems that are then tested against the project goals, budget, performance, and constructability issues identified in the previous steps. At this stage of the integrated design process, there will be some iterative loops required to go back and hold further study sessions to refine

some of the individual toolbox issues with the energy modeler, construction specialist, cost consultant, and maintenance people. McDonnell says, in this step, "the idea is that once the basic building design is assembled using everyone's individual 'toolbox' ideas, then the team starts refining the 'whole' design, to coordinate the design ideas and refine them. This phase chips the edges off everyone's individual ideas and fits them into a design whole."* This step carries forward into the construction documents phase.

7 During construction, the project team follows through with such steps as proper building commissioning, monitoring of building performance and maintenance issues, and establishing a post-occupancy evaluation program with the owner's operation and maintenance team.

This sequence of steps indicates how an engineer might view the process. What I find most interesting is the perspective that the integrated design process continues through the entire construction, start-up and first-year occupancy.

PLATINUM PROJECT PROFILE

Toronto and Region Conservation Restoration Services Centre, Vaughan, Ontario

Serving as office space for 36 occupants and housing a works garage, the Restoration Services Centre is an 11,700-square-foot, two-story facility. The Toronto and Region Conservation's Restoration Services Centre was designed to reduce energy costs by more than 66 percent compared to Canada's Model National Energy Code. The facility estimates a 57 percent reduction in energy consumption through a ground-source heat pump, radiant slab heating, energy and heat recovery ventilation, reduced lighting power density, and an improved building envelope. Composting toilets, water-free urinals, and low-flow lavatories contribute to an 80 percent reduction in potable water use.[†]

Integrated Design in Practice—
An Architect's Experience

Now that we've heard from the engineers, let's take a look at how some leading architectural practitioners of integrated design approach this task. Stephen Kieran is a principal at Kieran Timberlake Associates in Philadelphia. His firm worked on the new

---

*Geoff McDonnell, personal communication, April 26, 2008.

†Canada Green Building Council [online], http://my.cagbc.org/green_building_projects/leed_certified_buildings.php?id=80&press=1&draw_column=3:3:2, accessed April 2008.

Sculpture Building and Gallery at Yale University. Kieran describes how they went about creating an integrated design process.*

> We didn't have the time to mess around with [a more conventional approach]. These were half-day meetings and everybody sat in the room. What transpired was pretty remarkable because the team started to comment on each other's work. We had some of our best criticism about mechanical systems from our structural engineer for instance. It became a real broad vetting process. Ideas bubbled up to the surface and were more thoroughly integrated as a result of having everyone in the room.
>
> For example, the building has a displacement ventilation system. It's the first one Yale has done. It's completely married to the structure of the building. All of the ventilation cabinets, the outlets for this very low-velocity ventilation are built around all of the structural columns. We did that because for the long-term flexibility of the building, the structure is not going anywhere and the displacement ventilation system needs to become a fixture in the building. We didn't want to compromise the flexibility of the building from the owner's perspective; it's basically a loft building.
>
> The same circumstance repeated itself in hundreds of details throughout the building. It was basically unifying and integrating rather than dividing and segregating. We didn't even try to get to Platinum. The owner only authorized Silver. They didn't authorize paying for anything that could get us anywhere above Silver. By all of us working together and integrating our efforts, we got to Platinum without expending a nickel more than the owner thought they would spend to get to Silver. We think it was really a by-product of all of those integrated systems in the end. It didn't cost anymore; it was just more intelligent design.

What lessons can we learn from this project? First of all, recall from Chap. 1 that there was incredible time pressure; only 21 months from start to finish. Second, there was a strong bias toward creating a high-performance design. Third, the team was really first-rate. Fourth, the architect was determined to create a process for truly integrated design and gave each specialty the opportunity, even the mandate, to participate fully, even in areas where they didn't have particular expertise. With sufficient team building and the support of the team leader, seasoned building team professionals are able to help each other toward more efficient and sustainable design solutions. No one can work completely independently or with only minor coordination (aimed mostly at avoiding obvious conflicts) and create a truly integrated design. Instead, there has to be a strong overriding vision and clear sense that everyone's contribution may have value.

Looking at the final result, Kieran commented that his experience with an earlier LEED Platinum project had led him to believe that one always had to add systems

---

*Interview with Stephen Kieran, March 2008.

and cost to get to high performance. In this project, his experience was quite different.

> Based on the Yale experience at least, it was clear to me that you really have to modify the process at the outset. It's not just us as designers and consultants but also with owners. By the way, the builders were in on all of these sessions as well. They attended all of the bi-weekly design sessions and were providing cost information from the outset. The builder was a firm that wasn't necessarily noted for building high-performance buildings but they got into it and developed the whole process along with us. They were pricing it as we went.

> We just got there by making intelligent and original moves to mitigate how much work the building ever has to do. The north-south, long-axis orientation of this building could not be more perfect, given the use and program in it. The way we developed the building envelope really worked to mitigate how hard the building has to work to do anything. In retrospect, it was all about mitigating horsepower requirements and maximizing or optimizing what the natural world gives us. We really sought to start out in unison with nature rather than working against it. We were able to get 52 LEED points without doing a lot of extraordinary things, just by good design. We used a displacement ventilation system, which is a very high-performance, low energy use system to begin with, so it didn't have to work as hard. The systems often have to work less because of what we did with design—both the orientation and the building envelope itself. I now have a new point of view and no longer believe that you have to expend more money to get to LEED Platinum, and the proof is in this building.

International Integrated Design: The New York Times Building

Bruce Fowle led the design team at FXFOWLE Architects for the design of the New York Times building (Fig. 3.6) in New York working with Renzo Piano Building Workshop. Here's his story about how integrated design is practiced on a major new urban high-rise, with a strong collaboration between two very talented design firms.*

> The [New York Times Building] was a collaborative effort from start to finish. While Renzo Piano was definitely the visionary for the project, both firms had equal roles in carrying out the project's design and in the execution of its construction. It wasn't the classic designer/executive architect relationship, as it was a project type that Piano's office was unfamiliar with and could not effectively design without our expertise and local know-how.

---

*Interview with Bruce Fowle, February 2008.

Figure 3.6   Designed by Renzo Piano and FXFOWLE Architects, the New York Times building in Manhattan incorporates many energy efficient, sustainable, and high-performance features including an incredibly artistic external shading system. Working with the Lawrence Berkeley National Laboratory, The New York Times Company created a state-of-the-art dimmable lighting and shading system designed to reduce energy use by 30 percent. Forty percent of the power required for The Times space is generated by an onsite cogeneration plant.* *Image courtesy of FXFOWLE Architects, photography by David Sundberg/Esto.*

What was conceived in the original competition was basically a tower on a base, but it ultimately evolved into a ground-based tower with an attached four-story structure, as you can see in the built condition. The primary element that was retained from Renzo's original vision was the notion of having a very light, transparent building, and something, as he says, that would vibrate as it refracts light from the sky. On the outside are tubular ceramic screens, which are suspended outward from the basic glass box, adding

---

*Forest City Ratner Companies (November 19, 2007). "The New York Times Company Enters the 21st Century with a New Technologically Advanced and Environmentally Sensitive Headquarters." Press release. Retrieved on May 28, 2008. http://www.newyorktimesbuilding.com/.

a layer of complexity as well as a tremendous amount of richness to the building. These screens also provide sun shading which is a primary feature of the building's greenness [by reducing the air-conditioning load and subsequent electricity demand].

Early on in the design process we made a joint decision with the owners that we would meet every month, alternating between Renzo's office in Paris and our office in New York City. This worked out very well as we could get much more of Renzo's attention when we were in Paris. The rhythm of every month was good because we were able to do enough work between meetings to move the process forward. In the early stages, some of Renzo's people were in our office full time where we had workstations available to them. Occasionally, we would send members of our consultant's staff to Paris for extended periods. For the most part, half of the work was done in their office and half in our office.

The more vital design elements, like the development of the façade and the external character of the building, were done primarily in Paris. During the construction phase, there were usually one or two people from Renzo's office here helping with coordination, field changes, and clarification of the design intent. It was not a situation where the design architects turned it over to the local architect to implement the construction documents and oversee the construction administration. It was really a collaborative effort although the emphasis of the work during the construction phase was out of FXFOWLE's office and the emphasis during the early design was out of [Piano's] office.

Because the building design started in 2000, prior to the development of the USGBC's LEED rating system, it did not go through the LEED certification process. Nevertheless, the New York Times Building is a classic example of integrated design in a very demanding environment, using international cooperation between two leading architectural firms as the touchstone for the process. For example, in discussing the role of the key engineering consultants in this project, Bruce Fowle says:

The engineers were brought into the process right from the beginning. You can't really do a high-rise building like that without the engineering input since it has so much influence on the integrity and character of the building. This was particularly true in this case where we were exposing the structure, which was the first time that had been done in the U.S. (exposing structure on a high-rise building of this nature). It demanded a very integrated process.

The engineers participated in on all of the weekly team meetings and in the majority of monthly meetings with the owners, either in Paris or New York. Each week we would determine work plan for the following week. We have a system of writing meeting memorandums that tracks all open items and makes it clear what action is required from each party. So, for instance, the memo might read, "Flack+Kurtz is going to study the use of XYZ mechanical system." The complexity and numerous innovations of the project demanded many specific consultant meetings in addition to the regular meetings. We had 30 or 40 different consultants involved in this project. Other than providing the vision and design leadership, the architect's major role is to act as an orchestra leader for the process. Most of the consultants were local and familiar to us, as it was important that we know their capabilities and working styles, enabling us to wave the baton and say, "We need the 'piccolos' now, the 'drums' next, then followed by the 'violins'." It's very much a part of our normal role, but particularly so in this project because of its uniqueness and that it involved so many people.

Another international design project involved a new building at Northern Arizona University (NAU) in Flagstaff, Arizona. The smallest of the three state universities in the state, NAU was the first to commit to a LEED Platinum goal for its new Applied Research and Development (ARD) building. The designers were Hopkins Architects of London and Burns Wald-Hopkins Shambach Architects of Tucson, Arizona. Robin Shambach was the project architect for the Tucson firm. Here's her report on how integrated design helped this project achieve its high-performance goals.*

The design architect, Hopkins Architects Ltd. is a London firm; they worked with the London office of Arup. The whole team was involved in the project from the very beginning to the very end. In the beginning of the project, Hopkins Architects took the lead on the design side with the Arup engineers in the U.K. They started with site analysis and climatic research for the specific bio-climate of Flagstaff.

[Later], the emphasis moved from the design team to the executive team and that was our office and Arup, San Francisco. With Arup, it was a smooth transition to the San Francisco team. Hopkins Architects were still involved and collaborative because we were implementing their design.

The team looked at all of [the preliminary site studies] and then met with the engineering group to identify key strategies that were established very early on. Those strategies included things that impacted the architecture and the engineering; that's really the key point of integrated design. Those things were the use of thermal mass, passive solar orientation, low-velocity air distribution (an underfloor air distribution system), the integration of photovoltaic and other alternative energies (solar hot water) and introduction of daylighting. Those key concepts really drove the building design including the selection of the materials, the structural frame (cast in place concrete) and the orientation of the building (south, southeast in a very long thin form). The building section is critical to the success of the project, which includes a long, thin floorplate cut in sections. The three floors of occupied spaces are connected by a gallery. That gallery allows for the introduction of daylight from the south and southeast. The negative aspects of solar gain (in adding to air conditioning demand) were mitigated with shade structures integrated into the building. It allows that gallery space to be a tempered space by using the thermal mass of the structure. That was all driven by those key concepts that were established very early on with the engineering team. The building design really came out of that.

---

*Interview with Robin Shambach, Burns Wald-Hopkins Shambach Architects, February 2008.

Shambach talks about the importance of the owner's decision-making role in achieving LEED Platinum results and the compromises that had to be made in this project to achieve that goal.

> Because the goal of achieving LEED Platinum came so strongly from the university, they were very committed to reaching this goal. When things threatened that goal, they were willing to prioritize and give up things to keep the Platinum rating intact. They gave up building area to achieve a Platinum rating. The building was originally programmed to be 100,000 square feet and the final building was 60,000 square feet. They gave up area in order to achieve a Platinum level of quality.
>
> The key for managing costs is that everything had more than one use. The gallery is a good example. It does so many things for the building but you only pay for it once. It provides natural ventilation, daylighting, effective thermal mass and gathers the solar energy (in winter). The concrete was not painted but rather it was tinted with a light color. So it was not only the thermal mass, but also the finish material. We didn't have to pay for an additional ceiling below that. The color of it also helped maximize the light use in the building, providing a reflective surface. Minimizing finishes and having the building be the finish material was a way to control costs.

In this case, the overriding goal of having a LEED Platinum building led to some major design compromises, such as reducing the building area. Shambach says that the keys to sustainable design are "thinking about it from the moment you hit the site, keeping the goals in mind throughout, making sure everyone is aware of the choices that have to be made and integrating the whole team from the beginning."

PLATINUM PROJECT PROFILE

Applied Research & Development Facility, Northern Arizona University, Flagstaff, Arizona

Located at an elevation of 7000 feet, the Applied Research & Development Facility is a $22 million, 60,000-square-feet laboratory and office building that took 4 years to design and build. The building's passive ventilation, radiant heating and cooling, heat-recovery ventilation, high-performance glazing and solar shading were able to reduce energy use by 83 percent compared to a similar conventional building. A photovoltaic system produces about 20 percent of the building's electricity, and a solar thermal system supplies hot water. Water-free urinals and low-flush toilets using reclaimed water contribute to a 60 percent reduction in water use. Seventy-six percent (by cost) of the wood used in the building was certified to Forest Stewardship Council standards.*

Environmental Building News, www.buildinggreen.com, March 2008, page 7.

Photography by Timothy Hursley.

The Contractor's Role in Integrated Design

No discussion of integrated design would be complete without examining the contractor's role in the process, especially since many high-performance projects are design/build or design/assist projects in which the contractor is engaged right from the beginning, at the very least to provide pricing and constructability reviews as the design is evolving. It's hard to overstate the importance of the contractor's role in an integrated design process. After all, in most projects, they coordinate the work of dozens of trades and subcontractors and have the responsibility for spending more than 90 percent of the project budget. They are also directly responsible for more than ten points in the LEED-NC system. Indirectly, they are responsible for implementing the design team's vision and holding the project's specifications firm against a barrage of submittal (change) requests.

DPR Construction's Ted van der Linden provides some perspective on how the general contractor might be a significant asset to the integrated design team.*

---

*Interview with Ted van der Linden, DPR Construction, February 2008.

We're a design/build contractor on many projects, so we get very involved in the design. The old notion was an architect drew it, the engineer designed the structure and we would get a comprehensive package that asked for our price. That's changed completely in the green building arena. We have not done much of the "hard-bid" type of work. [For us and for most of these green building projects,] it's no longer the traditional design-bid-build process. It's a holistic design process where we're hired at the same time as the architect, in some cases even beforehand. We have a great opportunity at that time to bring up green, we might say, "Have you considered making this a green project?" The owner or the architect often say, "We've heard that costs more." I typically reply, "We've done 45 LEED projects and they all have been between one and two percent in first-cost premium. We don't know the design strategy yet but we can certainly help you down that path." Then we might hear, "That great, only one percent. That's a rounding error on a $40 million project."

In these cases, we get to bring our arsenal of products and materials that we've successfully used and suggest those items to the architects for inclusion in the specs. We don't take the liability of being the provider of the specifications for the projects. We do what I like to call greenlining. We go in and greenline the current specs and say things like, "These aren't low-flow toilets. This isn't a low-VOC product." We become the constructability reviewer.

We're doing a LEED Platinum project for a very popular sports-type company. They are really into organic and all things green. When we showed up at the table, it was one of the first times [for us] that the architect, the engineer, the whole project team was already at a very high level of green thinking. Yet, we had an existing building and we were going to deconstruct parts of it. Architects aren't generally used to taking things down and reusing the materials. We brought in a wood-certifier and had them evaluate the building. They provided a list of opportunities (for reuse of the materials) to the project team. The architect could then say, "We could use that reclaimed lumber here, now that we know the quality and availability." That's an example of where we might say, "We want to go after this credit. Let's talk about deconstructing this building and reusing it."

Sometimes the design team doesn't necessarily trust us—that we're in it for the right reasons and doing green for the right reasons. In our case, we're an open-book contractor. We share our numbers, our bid, our fee and our markup. We treat it like we want to be part of the family. It's the notion of being married to a client for a period of time. We still want to have a great relationship with them "once the kids go to school"—through the whole process.

From this perspective, you can see that a truly integrated design process will have to include the contractor as well as the future operators and occupants of the project. It's not really a high-performance design unless it can be built for close to a conventional budget and operated in a sustainable manner by the people who are actually going to be responsible for it. So, to be successful, integrated design has to take the long view of building construction and operations.

PLATINUM PROJECT PROFILE

Betty Irene Moore Natural Sciences Center, Oakland, California

The 26,000-square-feet Betty Irene Moore Natural Sciences Center at Mills College houses classrooms, laboratories, and research facilities. The total building cost was more than $17 million and the project was completed in July 2007. Designed to be 43 percent more energy efficient than California Title 24 requirements, the building is also 89 percent more energy efficient than a typical academic lab in the San Francisco Bay Area. A rooftop mounted 30-kW photovoltaic system is expected to produce over 46,000-kWh of electricity annually. The rainwater collection system will contribute to a 61 percent reduction in water use. The building design incorporates daylighting, operable windows, radiant slab heating, displacement ventilation, and evaporative cooling.*

Courtesy of Cesar Rubio Photography.

*Mills College [online], http://www.mills.edu/news/2007/newsarticle12112007platinum_award_LEED.php, accessed April 2008.

A New Trend—The Integrated Office?

It appears that some architects have decided that the best way to deliver integrated design is to have an integrated design team within the same office. While there are a large number of "A/E" (architect/engineer) and "E/A" (engineer/architect) offices among larger companies,* there appears to be a trend among a few firms to bring the key building designers all under one roof: architects and interior designers, structural and mechanical engineers, and landscape architects. California's LPA, Inc. is a midsized green architecture firm that adopted this approach. President Dan Heinfeld says:[†]

> My firm looks at the world pretty differently because while other people talk about integrated design, we actually changed the way our office practices. We have architects, engineers, landscape architects, and structural disciplines all in house so that the office is more integrated with people outside the traditional "design" disciplines. I believe it's profoundly different when those disciplines are in house. You really can look at integrated design when all of those disciplines are in balance trying to find the right sustainable solution.
>
> On our projects there is someone represented from each of the disciplines at our sustainable charrettes, which basically happen at the beginning of a project. It's not additive or subtractive, it becomes part of the building design's DNA.
>
> Internally, it's easy to get people involved in the [integrated design] process because we've been talking about this for a long time. It's being driven by two things: One, we know that this integrated approach will lead to better buildings and architecture and the second is that market forces want buildings delivered in different ways, whether that be design/build, integrated project management, or using Building Information Modeling (BIM). The market place is leading the discussions, "We want projects delivered differently. We want the design world to work differently."
>
> It is profoundly different when architects and engineers are sitting next to each other and having those spur-of-the-moment discussions involving not only design issues but also structural and mechanical issues. For example, an engineer, an architect and a landscape architect get together to talk about a project and look for that right sustainable solution as opposed to sending each other a drawing, setting up a meeting, etc. It becomes a much, much more integrated process which we know will lead to a more sustainable solution.

Whether this trend will become more widespread is open to question, but it is one increasingly common response to the shortage of consulting engineers willing to engage architects early in the building design process, so that the building team can produce a better integrated and higher-performance result. Portland, Oregon architect James Meyer is not convinced. He says, "I could argue all day that diverse trends brought together in a multidisciplinary team with members selected specifically for each project will result in a better project than one where the members happen to work for the same firm. The "best" do not likely live at the same firm when it comes to [forming] a full team."[‡]

---

*For example, A/E or E/A firms represented 19 of the top 50 (38 percent) "pure" designers in the annual survey of the top 500 design firms conducted by *Engineering News-Record*, April 21, 2008, p. 46.
[†]Interview with Dan Heinfeld, LPA, Inc., February 2008.
[‡]James Meyer, Opsis Architecture, personal communication, May 2008.

4

THE ECO-CHARRETTE

One of the key elements in the integrated design process is getting people together into a high-performance work setting. The eco-charrette is a means that adapts the well-known architectural design charrette specifically to the challenges of high-performance buildings and even more specifically to achieve high-level results using the LEED performance evaluation system. In my own practice, I prefer to have at least two separate events: the first is a visioning session that involves perhaps only the architect and the client's higher-level decision makers, with the specific task of crafting a vision for the project. Often I pose the question: "It's now 2018; looking back on this project from ten years' distance, what do you most want it to achieve, what do you most want to be proud of? What do you hope that the building occupants and other stakeholders will value most about this project?" Sometimes referred to as "backcasting," questions like these are designed to appeal to the higher-level emotions and values that most people bring to a major building project. They can serve as a guide to detailed design approaches and especially to making the inevitable tradeoffs that accompany any building project. Notice that none of these questions uses the word "LEED" or "green building."

Second, following the visioning session, which can usually be done in less than a day, I like to hold a detailed design charrette with a full range of building team participants: the architect, the owner's project manager, sometimes key stakeholders from the owner's side, the general contractor (if one has been chosen) and key consultants, including at least the mechanical/electrical engineer, lighting designer, structural engineer, and civil engineer. The landscape architect and interior designer can be added, depending on circumstances. This session should focus on design opportunities and should result in a clear plan of attack to resolve major uncertainties. I try to keep the LEED "scorecard" in the background. I find it useless to ask someone whether a project is within a quarter mile of two or more bus lines, for example, as called for in LEED-NC sustainable sites credit 4.1. It either is or it isn't, and that can usually easily be determined. So there's no sense in tying up $1000 per hour (or more) of design talent while someone figures that out. The LEED scorecard is a useful tracking tool during schematic design, but really doesn't belong in the eco-charrette. The subject matter of LEED of course does belong: site design, the water balance, energy conservation, renewable energy opportunities, waste management, and the desired level of indoor environmental quality. These are good subjects for discussion at early stage design charrettes and often yield impressive results when the entire team tackles them as a group. Let's take a look now at one such project and find out how a creative approach can generate new avenues for sustainable design.

PLATINUM PROJECT PROFILE

Banner Bank, Boise, Idaho

A 195,000-square-foot, 11-story Art Deco building in downtown Boise, the Banner Bank is a commercial Core and Shell office building. The total project cost was $25 million or $128 per square foot. The reduced operating costs are estimated to increase the value of the building by $1.47 million with a 32.4 percent return on the developer's incremental investment. A system that captures and

Alpha Image Photography by Giuseppe Saitta.

reuses stormwater from downtown Boise streets and parking lots, contributes a 60 to 80 percent reduction in water use compared to similar buildings. The facility uses 50 percent less energy for HVAC, hot water, and lighting loads compared to a typical office building of the same size including a 65 percent reduction in electricity used for lighting.*

The Charrette Process

Most observers agree that the key to the charrette process first involves listening, then participating creatively. The key to charrette facilitation is to create different avenues for participation, so that the group benefits from everyone's contribution (Fig. 4.1). Dan Heinfeld, president of LPA, Inc., a leading architecture firm, says:[†]

> Part of the charrette process lies in knowing that a good idea can come from anyplace. You have to be willing to accept that. It doesn't matter where the idea comes from. When those things happen, it's so clear. You can't bring a bunch of [solitary] egos to an all-stakeholder meeting because it just doesn't work. The other part is to be willing to let go [of control] and let the process run its course. If you really believe in the process, you'll believe in the outcome also.

> We're working on a student center for California State University, Northridge now. It's basically paid for by student money, so there are four students on the committee. In one of the early design meetings, one of the students said something like, "Why do these gyms always look the same?" Our team took that [question to heart], and now we're doing a project that doesn't look like any gym you've ever seen. It's remarkably different, because first, it's a sustainable solution and second, because it's the result of a truly integrated design process.

As you'll see in the stories below, without a strong charrette process, what gets designed tends to be pretty conventional and typically comes from the previous project experience of the architect and engineers. The problem is that few of those previous projects represent high-performance projects at the level most of us would like to see, that is, 50 percent or more energy savings, LEED Gold or Platinum certified, with a significantly reduced carbon footprint.

SWOT Analysis

My friend Nathan Good, an award-winning green architect in Salem, Oregon likes to use a SWOT analysis for his visioning sessions, using a diagram much like the one shown in Fig. 4.2. Borrowed from the world of the MBA school and from management

---

*FM Link [online], http://www.fmlink.com/ProfResources/Sustainability/Articles/article.cgi?USGBC:200703-19.html, accessed April 2008.
[†]Interview with Dan Heinfeld, LPA, Inc., February 2008.

Figure 4.1   (Left to right) John Nystedt, Muscoe Martin, Colin Franklin, Tony Aiello, and Bert Westcott, design team members of the Morris Arboretum project.

Photography by Paul W. Meyer of the Morris Arboretum, University of Pennsylvania.

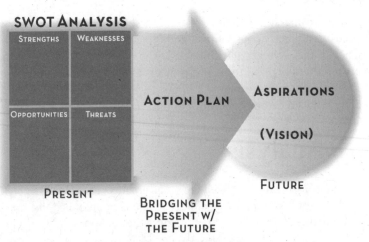

Figure 4.2   Vision-SWOT-Action describes a process for charrette facilitation that engages participants and generates a long list of design issues and tentative project decisions.

consulting, the SWOT analysis combines a candid discussion of the following four elements with a future vision, to arrive at an action agenda for the project. The SWOT analysis can be used to uncover opportunities, such as external funding for energy efficiency upgrades, things that might fall outside of any individual discipline's particular responsibilities. It provides a convenient way to create a bridge from the present situation to a desired future outcome.

- Strengths—internal to the project; these can be particular design talents, owner resources; and the like.
- Weaknesses—internal to the project; things that can inhibit sustainable design, such as disagreements over goals, lack of resources to pay for energy efficiency upgrades, and so on.
- Opportunities—external to the project; these can be natural resources (solar, wind, etc.), financial incentives for green buildings; partners willing to invest in the project, and the like.
- Threats—external to the project; these can include anything that would threaten the success of the design process, such as a change in ownership or local laws that do not allow graywater systems.

The University of Pennsylvania Morris Arboretum Project

Robert Shemwell is an architect and principal of Overland Partners, San Antonio, with a string of high-performance, sustainable design projects to his credit. Called upon to design a new Horticulture and Education Center for the Morris Arboretum for the University of Pennsylvania, Shemwell's team took a unique approach to the design charrette process.* As of March 2008, the project was in design development and aiming squarely at a LEED Platinum rating.

If you're going to have an integrated design process you have to bring everybody to the table at the initiation of the project. In this case, we had several daylong brainstorming sessions with the clients, board members, staff members, people from the university's facilities group, and all of our consultants. We started with a couple of different things. We went through an image creating process over a social hour, our first evening. At that time, we did two exercises (Fig. 4.3). We drew two copies of same object similar to a "paint by number" outline and cut them up into squares. We had two tables set up with pastels. At the first table they came to as they were coming into the dinner, for five minutes or so, they would sit down and color one of the squares. They could color it however they wanted to. Then they went to the second table where they would pick up a square, which included instructions about what colors went where.

---

*Interview with Robert Shemwell, Overland Partners, February 2008.

Figure 4.3   An exercise during the Morris Arboretum charrette invited participants to select photos to visually express "what we are," "what we are not," and
"where we want to go." *Photography by Paul W. Meyer of the Morris Arboretum, University of
Pennsylvania.*

Then we went behind the curtain while people were having some food and assembled
the two drawings. The first one was made up of all the squares the people colored
independently and the second one was all of the squares that people colored with
instruction. At the end of dinner, during desert, we brought those out and when we
held up the first one and said, "With everybody doing their own thing, this is what it
looks like." Then we held up the second one and said, "With everybody moving under
some direction, this is what it looks like." It took the same five minutes to do each
exercise but here you have a beautiful picture of the barn. In this case we used the old
Amish barn at the arboretum.

The first drawing was a surreal image with colors going everywhere that you could
sort of see might be a barn. Then the other one was actually a pretty good rendering
of the barn. We used that as an object lesson just to talk about process. In other words,
if we all pull together and work toward a common goal, in a relatively short time, we
can pull together something that begins to create a vision that you can understand and
is attractive. We used an object [the barn] that they have a great deal of affection for
to really get everybody thinking about the fact that this is all about working together.

We followed that up by walking the group through what we would be doing over the
next several days. We let people know what was coming, what was expected, when
they needed need to be there, what times were optional, why they needed to be there

and what outcome we were hoping to achieve for each piece. We really like to do a dinner before the brainstorming session because it brings a group of people together who don't know each other and just by the course of doing that little exercise together and eating together, they gain an understanding of what they're going to be doing over the next several days, and they come in the next day with their feet on the ground. They know what they're doing, where they're going, why it's important and with some excitement. It helps reduce a lot of the fear and trepidation.

I think this whole aspect is usually overlooked in the integrated process. People forget that people are people and they're highly relational. The first job that you have to do is build the armature of relationship and the armature of trust. So the first thing we did is focus on using that time to begin to build relationships, trust, and an understanding of what we're doing.

In this case, we started the next day with two exercises. The first exercise that we did was an image exercise. We had a huge pile of images taken from all kinds of references out on a table. We asked people to go through and pick three sets of images, they could pick as many as they wanted in three different categories: who you are, who you are not, and who you want to become as an institution. We had people who selected the images talk about them. A lot of times they will pick really odd images.

For example, they might pick a little kid sitting on a stump or a picture of the sky in a certain way. In this meeting someone picked an image of an Amish buggy going down the road and said, "This is who we're not. We're not old fashioned. We're not stuck in the past." You get some very interesting responses out of it. All of that is designed to give us clues as to what they're thinking. We have a saying, "The minute the architect picks up a pencil a whole bunch of people are cut out of the conversation." People are intimidated by those who can draw. But ideas and words are common currency.

We follow that up with a card session. We lay out a series of areas such as budget, schedule, image, environment, and any category that we think is important for that project. We hand out index cards for each category. We ask the group to write down their thoughts associated with the category listed on each card. We do this exercise because we found that if it's done verbally then the people who speak loudly and like to speak in groups rule the roost. The ideas of people who are a little more unsure, shy, or quiet don't get heard. We go through all of the cards and let people comment on what they were thinking and why.

An example might be the project's image. One response was, "Not ostentatious." Somebody else wrote, "Forward looking." Another response was a Quaker quote (they're in Quaker country up there), "Plain but of the best sort," which has actually become a very important part of the project. The idea is that it's simple and made out of good materials, but it's not about decoration. It's really about the function and quality of the materials and the way those things come together to make something that is not only useful but really beautiful. That idea has become a cornerstone of the project. The ultimate goal of that exercise is to write a mission statement using the words that came out of the cards. We say that is the "constitution" for the project. In other words, we define in words what we're trying to accomplish before we ever start drawing.

We then broke up into small groups and asked each group to write a draft of the mission statement. We asked them to think about, "What is this project all about, what does it mean, and what should it do?" We're not asking about the institution but about the project, "What does the project do for the institution to help it accomplish its mission?" We had each group report back and we took the best pieces that came out of the different groups and pushed them together. By the middle of the afternoon we had a mission statement that they had written themselves. That became the direction for the project and we had agreement on that from the group. Before we ever start drawing anything, we have to get agreement.

This process is the heart of the charrette. All of the stakeholders and people from the various disciplines are involved in this process. If you don't own the vision and if you don't feel like you helped shape it, you're just a passenger, not a pilot.

We found that this gives the team a lot of open-field running to do. We spent the next several days taking the information, working on the site analysis, and coming up with the beginning of schemes. We had intermediate meetings with the critical staff members; in them, we commented on what we were starting to find out, where we think we were beginning to go, and shared some ideas. We got their reaction and then we went back and worked some more, then stopped again and got their reaction and commentary. While we're doing this process, we document it in real time. We bring a group of people to the meeting who are solely responsible for taking photographs, scanning drawings, and putting them into slides as we're going along.

During the concluding two-day workshop, there were several things going on at once. For example, we had one team working on drawings and at the same time we had another team meeting with staff, talking about interpretive aspects of the project. We asked questions such as, "What do you want people to learn here? How are we going to accomplish that? How can we use the site assets? How can we use the building assets? How can we use the artifacts that you have to teach people a larger lesson? Where do you see the opportunities?" We were working on that while simultaneously another group was working on the schemes. All of that gets thrown into the PowerPoint slides.

The next night we have a concluding dinner. There is a presentation done by all members of the team. For example, the landscape architect talked about how this fits into the master plan, the site analysis, lessons learned from site, what the river is doing, what geology and topography are doing together. We had both board members and staff present the vision statement back to the larger board. We walked them through how the site analysis would form the building, what some of the sustainable goals were, how we were breaking down things like the envelope systems, the site and what we were starting to do with the major components (Fig. 4.4).

We got a lot done in an incredibly short amount of time. If, at the beginning of this process, if you asked the people at the Arboretum, they would have said they never thought it would have been possible. We basically came up with the concept at that point. We spent the rest of the time refining and improving the concept. There were some changes made where necessary but the basic concept, the organization of it on the site and the attitude of this very simple, but well-articulated building have stayed very constant.

Figure 4.4   Designed by Overland Partners, the Horticulture Center at the Morris Arboretum on the University of Pennsylvania campus is aiming for a LEED-NC Platinum certification. *©Courtesy of Overland Partners.*

We tell people that it's perfectly okay to have an agenda but it's not all right to have a hidden agenda. We're open to all agendas. They are necessary; in fact, we tell people that we are looking for conflict. A big chunk of what we're doing in those few days is looking for the points of tension: Where are the things where it almost seems impossible to do this and that? Where people want to say: it can do this or it can do that.

We say that the point of genius for a project is when it can do this and that, not this or that. Again, we tell people, that it's okay if they don't agree. In fact, it's important that you don't agree, so we can find out what the real issues are, get them out on the table and then find a place for resolution. Doing that takes away some of people's fear about being heard. A lot of times, they are fearful about voicing things they may not agree with. We really prod them and say, for example, "Paul has said this and that can make a lot of sense in certain ways, but what are the problems with that? What could be a complication? What might not work? How does that not fulfill the vision statement?" And that gives people an opening to say, "Well I was kind of thinking that it didn't really do this or that." That allowed us to discuss those points. It's less about conflict than about points of tension. We're asking, "What are the things that are pushing back on each other?"

What you are eventually trying to find out is: which thing is going to take precedence and how is it going to make allowances for the other functions of the building? The educational group wants the building to do certain things. But the people on the horticultural side were saying, "Wait a minute, that's going to interfere with my work."

Eventually we found that we were able to create a concept that is sort of a half of a figure eight arrangement, a giant S curve. The front end of S curve is the public court and the public faces that and then it flips around and there's the work court and all of the work functions face that. A huge point of conflict for the project was that the horticulture staff needs to move around with big machinery, dangerous materials and things that they don't want the public to a part of. However, they do love the idea that people get a glimpse in and can see the amount of work that goes into maintaining the place. That is horticulture, that's the process and that was an important aspect of the building. This building is about process.

But on the education side, others were saying, "We can be a very educational place, we can bring the public here, we can address the public's needs and give them a dressed up face, with a glimpse into what the [horticultural] process is." If that hadn't come to the surface, as a big point of tension, it would have been easy to skip past it instead of making it the heart of the concept of the project.

The benefits of this process are that we get further faster and we start the design process with a huge amount of buy-in and a lot of integration. The engineering, landscape architects, the person in charge of the parking lot, person in charge of catering were all there and letting us know what his or her needs are. Even if we haven't quite got it figured out, we can start to shape the response around making allowances for everyone's needs. When the drawing is started early on, it's very easy to forget pieces because you're trying to deal with big issues. You can leave out very important parts of the program.

When you're talking about a LEED Platinum project, you need every inch of everything working together. Because what you're trying to do is create a building that is a system where everything works together. We talk about strategic, systemic and sustainable. The building is strategic; it's not an end in of itself, it's something that helps people get closer to their mission. The natural systems, construction systems, budgetary systems and all of the other systems create the context of the project, so that the project can become a systemic response. In other words, the ideal Platinum building operates as an integrated system, not necessarily a series of isolated components.

Let's break down the approach used by Bob Shemwell and his team and look at some of the elements. First, the opening dinner with a team building and mind-bending exercise indicates to everyone that something special is happening with this project and that something creative and fun is about to take place. Second, key stakeholders are engaged in a very uplifting way to examine the project from many different viewpoints to come up with a mission statement that will guide the project and help resolve any conflicts or trade-offs. Then there is a hiatus while the design team and key client staff work on design concepts and look for potential conflicts that can be resolved in a "both/and" fashion rather than "either/or." In this way, true integration is achieved and each person's viewpoint is respected, even if it can't be fully incorporated into the design. Finally, there is a concluding dinner in which the design concept is presented to the full stakeholder group again, looking for buy-in to guide the design team from then on. Building systems emerge "strategic, systemic, and sustainable" from this process. In the case of this project, where the architectural team is from Texas and the client in Pennsylvania, it's very important to build trust and a mutually supportive working relationship before everyone returns to their respective workplaces.

PLATINUM PROJECT PROFILE

The Casey, Portland, Oregon

Developed by Gerding Edlen, designed by GBD Architects with Interface Engineering and built by Hoffman Construction, the Casey, a 16-story residential building in Portland, Oregon's Pearl District includes 61 residential units, 4200 square feet of ground-floor retail and underground parking. The project cost was $58 million. The condominiums were designed to be 50 percent more energy efficient than code. The project's high performance and green features include: roof-mounted photovoltaics, a green roof, onsite stormwater treatment, ventilation energy recovery units, operable windows, high-performance glazing and water-conserving fixtures.*

Photography by Michael Mathers.

*Gerding Edlen [online], http://www.gerdingedlen.com/project.php?id=22, accessed April 2008. "The Casey, Probably the Greenest Condos in US," Jetson Green, December 5, 2007 [online], http://www.jetsongreen.com/2007/12/the-casey-proba.html, accessed April 2008.

Adopt "Right Mind"

From the experience at the Arboretum, you can clearly see the value of an extended design charrette in solving more intractable site and program issues. People need time to reflect; that's perhaps the one critical problem-solving element that's really missing during periods of intense design activity. Without periods of reflection and casual interaction, it's often possible to overlook good design solutions in the rush to finish the charrette and get on with the "real work" of design. We still fight the notion that ideas are vague and unpleasantly slow to come at times, whereas drawings are solid, concrete, and can be produced on a clear timetable. Here's a bit of sage advice for our hurried times: don't just do something, sit there.

James Weiner is a Los Angeles–based architect with a similar philosophy and approach. He says:

> My experience is that most groups of professionals on a design team speak about integrated design but they really don't know what it is. Even on a high-level LEED project, there's still a very clear firewall between many of the disciplines that come together on a project. We try to break that down by gathering as many of the people involved with the project from the start, right through to occupancy and maintenance. We get everyone into a room at the project initiation and discuss the values, goals, and processes that we're going to be using in order to move the design process forward in a sensible fashion.*

Weiner sees that one of the most important tasks of the charrette and the charrette facilitator is to force a change in standard approaches.

> By and large, the project delivery process is pretty deeply embedded. Owners will develop programs without necessarily talking with the design team who are going to be delivering them. They certainly aren't conflicted about developing building programs without involving the contractors who will build the actual buildings. Architects will often start working without hiring all of the consultants that ultimately need to lend their expertise to a project. They'll often defer hiring the engineering disciplines until relatively late in the game.

Weiner's point of view is something of an amused skeptic who is also a very passionate advocate for green buildings and sustainable design. I find this typical of the leading green building consultants and eco-charrette facilitators. Reflecting his particular approach, Weiner says:

> People need to first come together as humans who happen to use buildings. They can share their personal values and then figure out what their technical contributions are going to be for the successful expression of their values. That first step is very hard for a lot of teams. They'll look at that and say, "Gee, this is kind of soft and fuzzy."

*Interview with James Weiner, March 2008.

That response reminds me of the story about the purchase of a bicycle shipped here from Japan. The instructions were translated from Japanese, and there are often some curious translations when you do that. The first step translated was, "Adopt right mind." If you think about that, how long does that take? Does it take even one minute?

"Adopt right mind" is a way of saying, "take a fresh look at the situation." That could be the motto for the entire integrated design process. This reminds me of the classic Zen saying, "in the expert's mind there are few possibilities; in the beginner's mind there are many." The trick is to balance the curiosity of a beginner with the tools and experience of the expert. An engineer might express it as "going back to first principles" and beginning to design from the ground up. What one finds is that the really best architectural and engineering designers have a way of suspending disbelief long enough to come up with some fresh approaches, then putting all of their skills to work on the project to bring the initial design idea—which might be just a sketch or outline of an idea—to fruition.

PLATINUM PROJECT PROFILE

Lake View Terrace Branch Library, Los Angeles, California

Located in the San Fernando Valley, Lake View Terrace Library is a branch library and multi-use facility for the city of Los Angeles. Completed in June 2003, the construction cost for the 10,700 square-foot library was $4.4 million. Over 40 percent more efficient than California energy standards, the building shell was constructed with high-mass concrete masonry units (CMU) with exterior insulation and employs a night-flush venting cooling strategy. Approximately 80 percent of the building is naturally ventilated with mechanically interlocked windows controlled by the building's energy management system. Nearly all of the glazing is shaded during operating hours while providing glare-free daylight throughout. A building-integrated photovoltaic system provides some of that shading as well as 15 percent of the building's energy requirements. Bioswales and other landscaping features reduce stormwater runoff by 25 percent.*

---

*AIA/COTE Top Ten Green Projects, AIA: The American Institute of Architects [online], http://www.aiatopten.org/hpb/overview.cfm?ProjectID=289, accessed April 2008.

5

BARRIERS TO HIGH-PERFORMANCE BUILDINGS: WHY SOME PROJECTS SUCCEED AND OTHERS FAIL

Let's take a short pause and look at what might influence the successful undertaking of a LEED project. As of June 2008, there were more than 53,000 LEED Accredited Professionals (APs), so one would expect that the success rate (defined as eventual certification of a LEED registered project) would be increasing over time. It's hard to say, because there were still only about 1170 LEED-NC and LEED-CS certified projects, as of June 2008. That means that most LEED APs have yet to complete their first LEED-certified project and therefore have little "hands on" experience with the entire process, from start to finish. And, as many professionals know, design and construction projects are messy affairs, often with significant obstacles in the way of realizing the initial intentions of the owners and designers. Nonetheless, buildings do get built and occupied eventually. This book demonstrates a process that works for high-performing (LEED Platinum) projects and shows you how to apply this same process to your own projects.

As of June 2008, there were nearly 7200 LEED-NC projects registered for eventual certification, and slightly less than 1100 certified projects, about 15 percent of the total registrations. In all other rating systems, there were 4060 projects registered, and 435 certified, less than 11 percent. Look at the numbers in Table 5.1, comparing registered versus certified LEED-NC projects to date.*

These numbers are not precise, because some version 2.0 projects switched to become 2.1 (when that became available) and some version 2.1 projects switched to become 2.2 projects in 2005 and 2006. In addition, LEED allows some combining of version 2.1 and 2.2 credits. Nonetheless, they are indicative of one relatively significant problem: even allowing for the time lag in completing large commercial projects, 18 to 36 months after registration, *many projects that start out with the best intentions don't end up certifying.*

---

*USGBC Staff Data, prepared monthly.

TABLE 5.1 LEED-NC REGISTERED VS. CERTIFIED PROJECTS, MARCH 2008 DATA*

LEED-NC VERSION[†]	REGISTERED[‡]	CERTIFIED[§]	PERCENTAGE[¶]
2.0 (ended 2002)	624	238	38%
2.1 (ended 2005)	2134	352	17
2.2 (began end 2005)	3684	467	13
Totals	**6442**	**980**	**15%**

*Data furnished by USGBC staff to the author; the author assumes these are generally accurate.

[†]LEED version 2.0 generally applied to projects registered before the end of 2002; LEED version 2.1 applies to projects registered generally between 2003 and the end of 2005; LEED version 2.2 has been in effect for all project registrations since January 1, 2006.

[‡]Registrations through the end of 2002 approximate the number of LEED-NC version 2.0 projects registered. New project registrations through the end of 2005 approximate the number of LEED-NC version 2.1 projects registered. About 16 months have passed since projects were able to register under version 2.1 (that is, since the end of 2005). LEED-NC version 2.2 registered projects are generally those registered since the beginning of 2006.

[§]Most LEED-NC version 2.0 projects that are going to certify have done so. Many of the LEED-NC version 2.1 registered projects are still working on certification documentation. Some may still be in design or construction, owing to various delays, or their large size.

[¶]Percentage certified through the end of March 2008. Not meaningful for LEED-NC version 2.2 projects, since projects that registered in 2006 and 2007 are mostly still in progress.

This problem is akin to breaking off an engagement before marriage; there's embarrassment, but life goes on. In the case of LEED, if we're aiming to produce better buildings because the future of the planet depends on your efforts, it makes sense to find out why projects that start out with the best intentions don't wind up with a certification package.

Without close attention to the numbers, any company will tell you, real change is impossible to achieve and deviations from desired performance are impossible to correct. LEED's success as a catalyst to institutionalize innovations in the building industry depends critically on a realistic assessment of progress.* To date, one must judge the progress as mixed, based on the numbers in Table 5.1.

Let's look more closely at the results: With LEED-NC comprising more than 66 percent of all project registrations and 74 percent of all certifications to date (and likely to continue that dominant role), it's the most important system to analyze. Why is it that some projects do not complete the journey from registration to certification? It cannot be cost, or cost alone. In the 2006 study, *Greening America's Schools*, Greg Kats studied 30 LEED certified school projects and came up with an extra capital cost of 2% ($3.00 on an average base cost of $150 per square foot).[†] With most projects

---

*See for example, Matthew May's 2007 book, The Elegant Solution: Toyota's Formula for Mastering Innovation (New York: Free Press).

[†]Greg Katz, Greening America's Schools, 2006, available at www.cap-e.com (complete citation).

carrying a contingency of 5 percent, this cost increase is well within the contingency budget. And, if an average elementary school costs $15 to $20 million (100,000 to 133,000 square feet in size), even "typical" LEED certification/paperwork costs of $100,000 (counting commissioning and energy modeling) would only add 0.5 to 0.7 percent to the project budget.

I believe that social factors, much more than the perceived cost or difficulty of LEED certification, are at work. Three come immediately to mind:

1 LEED projects achieve the best results when the design and construction teams use an *integrated design process*. I believe and have observed in a variety of project meetings that most architects don't know how to manage such a process, are uncomfortable with it, or are unwilling to use it, and that most owners don't push their design and construction teams to change the traditional "serial hand-off" process.

2 Responsibility for LEED documentation is often not clearly assigned to a specific "LEED Project Manager" except when there is a LEED consultant involved. Even in that case, the consultant has little authority to demand documentation on a timely basis from the design team. If the documentation management is kept in-house, it's often assigned to a relatively junior person, again with little authority to command performance from other members of the building team. The solution here is to get some good LEED project management software in place, beyond the USGBC's project data templates, something that I present later in this book, and then require teams to achieve a certain specified results by using it.

3 Lack of integrated design training for all design and construction team members. Mechanical engineers, in particular, should step up and demand that they receive the same training as the other members of the project team.

If the problem is really about process, then building teams must become process champions and should look to other industries where process is deemed highly important, such as the Toyota production system in auto manufacturing. Laura Lesniewski, is a principal at BNIM Architects in Kansas City, Missouri and led the design team for the LEED Platinum-certified Lewis & Clark State Office Building in Jefferson City, Missouri. Concerning how to run a successful project, she says:*

I focus on two things: having the right process and the right people. In people, I would emphasize that the most important entity to be on board with the goals of the project is the ultimately owner of the facility. Then, ideally, the rest of the design team has to be on board. If you can get consultant team members that are at the top of their game, then that's all the better. Including as many people as early as possible is also important. If the delivery method is such that you can have the contractor at the table early, bring them on board, because they can provide invaluable advice, as you're moving through the process, in terms of helping with constructability and input on cost and schedule. For example, by tracking closely with the design team's efforts, they can tell you how design decisions are influencing cost and schedule in smaller batches of

*Interview with Laura Lesniewski, BNIM Architects, March 2008.

information and cost sharing, so that there are not major milestones in the project where costs are revealed and rework is required. Making minor course corrections along the way is much less painful that finishing a phase of the work only to find out you need to rework the design to take 20 percent out of the budget.

With regard to process, it is critical to provide clarity for the team in how you're going to get everybody integrated and at the table at the right time. Clarity on the decision-making structure and schedule is also critical so that everybody knows when they need to have something ready in order to hand it off to the next group to do their job. Another aspect of "lean thinking" is the potential for all team members to make "reliable promises" to each other. A simplified example may be to ask somebody, "I'm going to need this from you, how much time do you need for that?" They may say two weeks. I might say, "Do you really need two weeks?" "Well, I could get it to you in one week, but only if I knew that I would have this information from this other person." So you go to that other person and see if they can reliably promise that they can get that on schedule. This is closely tied to "pull" scheduling, which attempts to identify what is really needed at each phase or step in the process. It's especially helpful on projects that have a short-time schedule and where costs are an issue.

In order to understand the benefits of reliable promising, it is particularly helpful to track the promises kept and not kept in order for people to understand how well they are doing. This is not to be punitive, but to learn. Being engaged with team members that you know and trust makes this a very powerful and effective component of the project, and helps everyone to promise better and rely on each other more.

PLATINUM PROJECT PROFILE

Lewis & Clark State Office Building, Jefferson City, Missouri

Constructed on the site of the former Jefferson City Correction Facility, the Lewis & Clark State Office Building houses approximately 400 Missouri Department of Natural Resources employees. The cost for the 120,000 square-feet building was approximately $17 million. Designed to reduce energy consumption by 60 percent over a standard building, this project by BNIM Architects uses daylighting technologies, advanced electrical and lighting control systems, an efficient building envelope and highly integrated and innovative HVAC systems. Rainwater from the roof is captured in a 50,000-gallon storage tank and used for toilet flushing. Bioswales, drain tiles, and a native ecosystem along with detention ponds eliminate the remaining stormwater runoff. Photovoltaics supply 2.5 percent of the building's energy needs, and a solar thermal system supplies hot water.*

*Lewis & Clark State Office Building Earns LEED Platinum Certification [online], www.oa.mo.gov/purch/recycling/success.pdf, accessed April 2008. BNIM Architects and Missouri Department of Natural Resources Receive LEED Platinum Certification for Lewis and Clark State Office Building [online], http://www.bnim.com/fmi/xsl/press/archive/index.xsl?-token.arid=47, accessed April 2008.

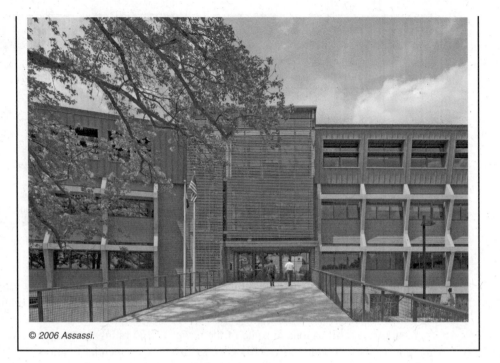

© 2006 Assassi.

Fewer Higher-Level Certifications

Another indicator of system difficulty is that the percentage of projects receiving basic Gold and Platinum ratings actually *decreased* from LEED-NC version 2.0 to version 2.1. (There are still too few data on LEED version 2.2 certifications to see if this trend is reversing.) This may mean two things: one is that more teams are completing their first project and haven't built the skill level to get to the higher levels of certification; however, it may also be that the costs of certification at higher levels are not coming down, leading teams on limited budgets to settle for lower levels of certification. Another reason may lie in the very success of the LEED system in selling its label to the marketplace. If any level of LEED certification is seen as a "green" building, equal to all others in the minds of the media and various stakeholders; then the actual certification level doesn't matter as much as the LEED label itself. In that case, project teams and owners may well think: Why spend more money than I need to get beyond basic certification?

That attitude may not be a bad thing. Even basic Certified projects are better projects than the vast majority of those that don't bother to certify at all. One study of 125 certified projects found that even the Certified projects were projecting energy and water savings of 30 percent versus a standard building. (See Table 5.2) Interestingly, the average water savings didn't increase much, even at higher levels of certification, whereas average energy savings increased from 25 percent to about 45 percent,

TABLE 5.2 PROJECTED RESOURCE SAVINGS IN 125 LEED-NC CERTIFIED PROJECTS

CERTIFICATION LEVEL (NO. OF PROJECTS)	AVERAGE WATER SAVINGS	AVERAGE ENERGY SAVINGS*
Certified	30%	25%
Silver	30%	25%
Gold	30%	44%
Platinum	30%	44%

*Vs. U.S. Department of Energy 2003 Commercial Building Energy Consumption Survey. The full study, "Energy Performance of LEED-NC Buildings," can be accessed at www.newbuildings.org.

compared with the reference building. However, if the goal of LEED is now to reduce a building's carbon dioxide emissions (largely tied to energy use in the building) by 50 percent, then even Gold projects fall a little short. However, the study found that 25 percent of all projects had savings in excess of 50 percent.

What about certifications in some of the newer LEED rating systems, such as LEED for Core and Shell buildings (LEED-CS), LEED for Commercial Interiors (LEED-CI), and LEED for Existing Buildings? Table 5.3 shows similar results for them. The pilot (beta test) versions of each rating system managed to certify between 25 and 30 percent of the projects, but the results for the "version 2.0" systems are disappointing at this stage.

What do this numbers suggest, for the other LEED rating systems? Let's look at them in turn.

LEED FOR CORE AND SHELL

First, LEED for Core and Shell is primarily used in the commercial speculative office market,* typically by large and fairly sophisticated developers; second, by using a LEED pre-certification for marketing purposes, a developer is pretty much "locked in" by tenant expectations and lease provisions to producing a project certification at the end of the project, so one would expect the percentages to be higher. Because these projects tend to be quite a bit larger than the typical LEED-NC project (average a 350,000 square feet vs. 110,000 square feet[†]), they take longer to build, and so many of the version 2.0 projects just aren't finished (and certified) yet. Additionally, at a 350,000-square-foot average, the typical project will cost more than $50 million, so the "fixed costs" of LEED certification are fairly small. Therefore, because of the marketing advantages LEED-CS confers on a speculative development project, it should

---

*To use LEED-CS, a developer must finish no more than 50% of the tenant space upon project completion; this would typically be done for an anchor tenant. Otherwise, LEED-NC must be used.
[†]USGBC LEED project data, April 2007.

TABLE 5.3 REGISTRATIONS AND CERTIFICATIONS FOR OTHER LEED RATING SYSTEMS[§§]

RATING SYSTEM	REGISTERED*	CERTIFIED	PERCENTAGE
LEED-CS v. 1.0 (pilot)	110	28	25%
LEED-CS v. 2.0[†]	1231	31	3
LEED-CI v. 1.0 (pilot)	106	30[‡]	28
LEED-CI v. 2.0[§]	895[¶]	186	21
LEED-EB v. 1.0 (pilot)**	88	26	30
LEED-EB v. 2.0[††]	891[‡‡]	44	5
Totals	**1120**	**203**	**18**

*Based on LEED registered projects listed in the USGBC web site, accessed May 30, 2007. These totals are lower than those in the "LEED Metrics" list.

[†]LEED-CS v. 2.0 became effective in the spring of 2006, so it is only about 2 years old, not enough time for most large projects to finish construction and certification.

[‡]Cumulative total as of mid-2007.

[§]In effect since October 2004.

[¶]Cumulative total as of mid-2005.

**End of 2004 registered projects total.

[††]In effect since October 2004.

[‡‡]Cumulative total as of mid-2005.

[§§]As of April 2008, data courtesy of USGBC.

have a strong incentive to proceed all the way from registration to certification. However, in examining the "percentage completion" rate in Table 5.3, we see it's no greater than for the other LEED systems.

LEED FOR COMMERCIAL INTERIORS

LEED-CI represents a wholly different situation, with an average size of only 50,000 square feet (one or two floors of an office building) and a quick project turnaround time. Most remodeling and tenant improvement projects happen quite fast, so that a decision to pursue LEED-CI must be made at or near the outset of a project. In that case, one expects the client and the design team to be committed to the end results. However, a countervailing factor is that the average project cost is less than $5 million, often less than $2 million, so that the fixed cost of completing a LEED-CI project (perhaps $50,000 to $100,000) is a more significant factor in the overall project budget. As a result, the percentage of registered projects already completed under LEED-CI version 2.0 (in place since November of 2004) is less than that of LEED for New Construction. As an example, Exelon Corporation, a major electric utility, used LEED-CI to fit out its new headquarters office space (Fig. 5.1).

Figure 5.1   Located in downtown Chicago, the Exelon headquarters contains 220,000 square feet of LEED-CI-certified office space. *Courtesy of Exelon Corp.*

LEED FOR EXISTING BUILDINGS

LEED-EB has been one of the slower programs to get going, because it involves benchmarking facility management and operations practices, often requiring a company to spend $50,000 or more (for internal labor or consultants) for the LEED effort as well as a few hundred thousand dollars for the upgrades required for certification in most cases. To date, less than 10 percent of registered projects (counting both the pilot version and the version 2.0, which has been around since November 2004) have received certification. Through March of 2008, corporate or for-profit entities had registered 62 percent of such projects. One expects the numbers of completed projects to be higher because a facility manager typically wouldn't register a project without having the funds and the commitment to complete the certification. Corporate life generally doesn't reward those who don't deliver, so the motivation for completing projects is pretty significant. Nonetheless, LEED for Existing Buildings' projects currently have a worse completion record than LEED for Commercial Interiors or LEED for New Construction. A countervailing factor that has emerged in 2007 and 2008 is the acceptance of LEED-EB as a platform for demonstrating sustainability concerns by some very large real estate management firms. For example, CB Richard Ellis, the world's largest property management organization, expects to submit more than 200 projects for LEED-EB certification by the end of 2008.* It is quite possible that LEED-EB is near the "take off" point in terms of market acceptance, a development that would be very useful for the emerging priority among many organizations to reduce their carbon footprints.

What Needs to Happen

What are the obvious fixes for the situation where most LEED-registered projects don't get certified? After all, if the USGBC's mission is to transform the building industry, the first step in completing the assignment is to make sure that those projects which start the journey actually get to the end of the road.

To get a better handle on what can be done, we interviewed several experienced LEED project management consultants around the country. Here are some of the fixes they recommended:

1 Have a realistic expectation of the costs of completing a LEED-certified project, both in terms of "soft" costs (design, process management, and documentation) and "hard" costs of additional net capital expenditures. Make sure that the project budget contains money for potential capital cost increases and, equally importantly, for LEED documentation.

2 Find a *sensei*, a master teacher (aka *consultant*) who can guide you through (at least) your first two projects, until you can clearly master all the technical, process, and documentation steps and until a "process champion" emerges from, or is assigned to the design and construction team.

3 Develop or purchase proprietary tools for LEED process management (see Chap. 8 for an example). It's different enough from conventional project design and construction to warrant a fresh approach. Make sure that these tools are employed on every project. Don't have people "winging it" by starting over each time a new person is put in charge of the LEED certification effort.

4 Building teams shouldn't expect to get paid extra for each project to do the same level of effort in green design and certification. My advice to them: get your costs down with each subsequent project. Look at LEED project management as a process amenable to the same process improvement steps as project design and delivery. One engineer noted the architecture and engineering firms "should be charging lump-sum fees for the value they bring" to an integrated design team.[†]

5 Project teams should take advantage of the distinction between "design" and "construction" credits in LEED-NC version 2.0 and submit the design credits for review as soon as construction starts. That way, they'll have a better idea where a given project stands (and will have assembled the documentation for the design credits) long before construction completion, while there's still time to add construction credits to reach desired certification levels.

---

*Personal communication, Warren Whitehead, CB Richard Ellis, December 2007.
[†]Paul Schwer, PAE Consulting Engineers, personal communication, May 2008.

6 Some people we interviewed said that there are institutional clients who are beginning to express the opinion that once they've designed and certified a couple of projects, they don't have to do it anymore, because they've proven that their design and construction process really is very green. This attitude, akin to "been there, done that," is a sure recipe for backsliding to the "pre-LEED" world. An owner without a responsibility to third-party documentation and certification doesn't demonstrate commitment to achieving higher-performance results and probably won't get them.

7 Owners and developers should hire those building teams with the most LEED experience if they want the best results. While obvious to many, this conclusion is often at odds with the tendencies of institutional owners to hire the teams they know best from previous (non-LEED) projects.

PLATINUM PROJECT PROFILE

Standard Refrigeration Company, San Juan, Puerto Rico

This two-story, 19,500-square-foot building serves as the headquarters for the Standard Refrigeration Company. Compared to a building built to ASHRAE 90.1-1999 standards, Standard's building was designed save over 70 percent in electricity costs. An 85 percent efficient enthalpy wheel capable of handling 125 percent of the required outside air is a major component in the heat recovery system. Standard estimates that the electrical cost savings alone will pay for the cost of the new building in 10 to 15 years based on today's electricity rates.*

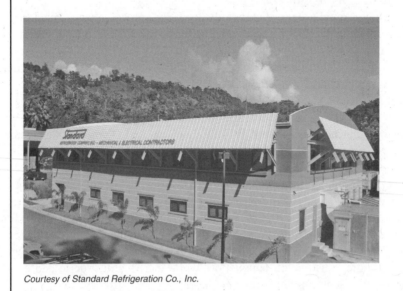

Courtesy of Standard Refrigeration Co., Inc.

---

*http://envirotechnews.blogspot.com/2008/01/greening-caribbean.html, accessed April 2008.

Finally, I think there's a message here for architects. As the "captain of the ship" for project teams, they have to take full responsibility if the voyage isn't completed, or if some of the cargo or deckhands are missing when the ship pulls into port. It's time for architects (in general) to get down to the serious business of getting high-performance green buildings actually built and certified. Isn't it time for the design professions to admit that they still have a hard time producing really green buildings each time out, every time, on or near conventional budgets? In my view, that admission of failure would open the door for real learning to start, and for the most part, architects need to lead the charge.

Getting Consistent Results

Let's take a look at how one owner, the University of Washington, tackles the issue of getting LEED certification on every project. Clara Simon was hired a few years ago as the first sustainability director for the Capital Projects Office. She relates her experience with LEED projects.*

Currently most of our LEED projects are state-funded projects. In the state of Washington there is a requirement that we need to achieve at least LEED Silver on all of our projects. (Through early 2008, the university had achieved four LEED certified buildings—including one LEED-CS Gold—with 14 LEED projects in process.) State funding drives the process a lot here on these projects, and sustainability is a key component.

We know at the very beginning that we're going to need to meet certain sustainability requirements so we begin with our requests for information (RFI) and requests for proposals (RFP); we're required to post ads in the paper about our projects because we're a public institution. We put right in there at the beginning that we need to meet at least LEED Silver. That's a trigger point at the very beginning when we're hiring professionals in terms of the direction that we're going to take on these projects.

The university recognizes that the key to the success of LEED projects is to streamline the process. We're constantly taking knowledge gained on each project and putting it into action for a higher level of success on the next project. The University of Washington has developed integrated design guidelines that cover a multitude of design choices for each project.

To implement the sustainable design process, I get on board at the very beginning in predesign and I meet with the architect and the engineer [who can be different from one project to another and typically are different]. We develop a building committee that usually includes the client, the architects, engineers, and even the landscape architects at times. There's a state law that says the university cannot bring the contractor on in that point in time, which is unfortunate; however, we can bring the contractor on board under a certain type of contract at schematic design. We try to develop our team up front. Usually I will give at least one presentation to the building team about what is sustainable design for us.

*Interview with Clara Simon, University of Washington, February 2008.

Our goal is to follow an integrative design process here. The state requires that we do an eco-charrette on all of our projects. Usually, that is a two-hour session. We used to do a full day or two days. We've streamlined it because we have a pre-meeting with the architect and the project manager to really determine the low-hanging fruit. So we don't have to handle that in the eco-charrette. What we handle in the eco-charrette is really high-level decision-making and problem solving about where we need to get to, to meet our sustainable directives.

One of the things we've recently implemented is an energy reduction goal in these buildings up front, for example, with the new student union building (Fig. 5.2). We use that as a driving force for the design. In the past, we would design it and then at the end we would look at where we were at [in terms of energy savings]. Now we've moved that analysis closer to the beginning of the project. In the integrated design process, I work with the environmental stewardship advisory committee here on campus. That's a committee that was developed under [University of Washington] President Mark Emmert. We're looking at sustainability totally on campus.

Figure 5.2   Designed by Perkins+Will, the renovated Husky Union Building on the University of Washington campus will house offices and meeting space for student activities, auditorium and performance spaces, a book store, bike shop, bowling ally, billiards, conference rooms, food services, banquet rooms, and an active outdoor area for food vendors. *Courtesy of Perkins+Will.*

The lesson learned by the university is that if you want LEED results consistently, you have to advertise that fact from the beginning. In addition, you need a dedicated person to oversee and drive the process. Finally you have to train not only your staff but also the contractors. For example, Simon and the university's team make an extensive effort to capture lessons learned and apply them to each new project. Also, architects are required to submit LEED documentation on a timely basis. According to Simon:

> Prior to construction, the architect is required to submit to the owner a list of specified products indicating the LEED performance reasoning, and the list needs to add up to the certification goals. Throughout construction, this document is used as a touchstone during the submittal and request for information (RFI) processes. Monthly LEED construction meetings are required to evaluate credit achievement status levels.

At the university's capital projects office, according to Simon, there are nearly 50 LEED Accredited Professionals. In other words, LEED certification becomes part of the culture of the place. Training contractors is also somewhat unique to the university. Simon explains how this is done.

> The state recognized that there were a lot of contractors that do not understand the LEED process. The training is done through the Washington Department of General Administration's sustainability office managed by Stuart Simpson. He put together a program called "Build it LEED" and worked with the Cascadia Green Building Council to set all of this up—the group that put together the Build It LEED training. I require that contractors on all of our projects follow that process.

Requiring contractor training as a condition for bidding or working on state projects is a process that every university, school district, local government, and state agency could follow. I'm sure that this training alone would lead to a significant increase in the success rate of LEED projects.

What's also interesting is that the process continues after occupancy. As Simon notes,

> After occupancy, the State requires ongoing metering and reporting of water, electricity and gas usage through 2016, along with building commissioning, and a post-occupancy evaluation at the end of 12 months. The State's goal is to benchmark each project with operating data, complete a case study and achieve an ENERGY STAR-rated building for each State-funded project.

6

THE BUSINESS CASE FOR
GREEN BUILDINGS

I often state the business case for commercial green buildings in 2008 in this way: if your next project is not a green building, one that's certified by an accepted national third-party rating system, it will be functionally outdated the day it opens for business and very likely to underperform in the market as time passes.* One expert claims that trillions of dollars of commercial property around the world will drop in value because green buildings are becoming mainstream and will soon render those properties obsolete.† This is a remarkable change in the environment for commercial green buildings in just a short period of time, beginning about 2006. The question for commercial developers is no longer whether they're going to build a certified green building, but when and where will they do their next projects, what level of certification will they seek, and how much they're prepared to invest in their green building program.

A recent study of commercial projects from the largest commercial property database proves this contention. Completed in March 2008 by the CoStar Group, which analyzed more than 1300 LEED-certified and ENERGY STAR buildings (of which 355 were LEED-certified) representing roughly 351 million square feet in its database, the study showed that not only are rents and occupancy rates higher in green buildings, but sales prices are substantially higher.

In the analysis, LEED buildings achieved rent premiums of $11.33 per square foot over traditional office buildings and exhibited 4.1 percent higher occupancy rates.

---

*There are buildings that may have green elements, but do not pursue formal certification. My estimate is that these represent less than half the green building market at present and will decline rapidly over the next three years as a share of all green buildings. As is often said, "The road to hell is paved with good intentions." The case for certification of buildings is made elsewhere. When people tell me they have a green building, but didn't bother to formally certify it, I say "prove it." The fact is that most people who claim to be doing green design but don't bother to certify the project through an independent third-party are likely practicing self-deception, since without certification as a goal, many of the green elements end up being cut from most projects for budget reasons.
†Charles Lockwood, "As Green as the Grass Outside," *Barron's*, December 25, 2006, http://online.barrons.com/article/SB116683352907658186.html?mod=9_0031_b_this_weeks_magazine_main, accessed March 6, 2007.

ENERGY STAR buildings commanded rent premiums of $2.40 per square foot and showed occupancy rates of 3.6 percent over peers.

LEED-certified buildings that were resold received $171 more per square foot, while ENERGY STAR buildings sold for an average of $61 per square foot more than comparable non-ENERGY STAR buildings.

LEED-certified buildings exhibited rent premiums of $11.33 per square foot over similar non-LEED-certified buildings and had 4.1 percent higher occupancy. Rental rates in ENERGY STAR buildings represented a $2.40 per square foot premium over comparable non-ENERGY STAR buildings and had 3.6 percent higher occupancy.

The data are now in, and they are compelling. This study demonstrated that green buildings achieve higher rents and higher occupancy rates, they have lower operating costs, and they get higher resale prices. What more information does one need, to make the business case for green buildings in the commercial sector?*

For institutional building owners, such as large corporations, universities, schools, nonprofits, and government agencies, typically those with long-term owner/occupant perspectives, the business case is even more compelling, since they can secure all of the benefits of building ownership, whereas commercial developers must typically split those benefits with tenants. Public and institutional owners increasingly see green buildings as expressions of policy choices favoring sustainability, a tangible expression of an often intangible commitment.

By 2010, most observers expect the business case for green buildings to be completely accepted and become part of a new definition of "business as usual." Richard Cook, AIA, a prominent architect in New York City (see discussion of the Bank of America Tower at One Bryant Park in Chap. 2), says, "In five years, it will be clear that buildings not reaching the highest standard of sustainability will become obsolete."[†] Many real estate professionals see the day soon when an office building will not be considered as "Class A" without a LEED certification. Without certification, buildings will be at a considerable disadvantage in most large cities, unable to command the same rents or the same quality of tenants as their certified competitors.

Here's the business case as expressed by Denver-based developer Aardex LLC's Ben Weeks.[‡]

Typically, businesses will pay anywhere from $300 to $600 per square foot for their people (assuming a salary and benefits cost of $60,000 per year and 100 to 200 square-feet of space per employee). And they'll pay $15 to $20 per square foot for rent. If a building can increase the productivity of the people by even 5 percent, well, at $600 a foot per employee that's $30, roughly twice or significantly more than the total cost of the building.

I think the sky is the limit for realization and maximization of with this relationship. We're sharing as much of this in the industry as we can, to anybody that will listen. For

---

*http://www.costar.com/News/Article.aspx?id=D968F1E0DCF73712B03A099E0E99C679, accessed May 11, 2008.
†Interview with Richard Cook, Cook + Fox Architects, New York City, March 2007.
‡Interview with Ben Weeks, Aardex, LLC, March 2008.

building and development, all costs are going up—the replacement cost of buildings, the cost of operations, and cost of hiring people—all costs are going up. As a society, we simply can't afford to, nor should we be willing to, accept the premise that buildings are disposable. Buildings have to be as productive and sustainable for as they can possibly be. At a minimum, they need to have a 100-year usable life or more. We believe that's not only quite possible, but *essential* to our industry, our communities, and the world.

Incentives and Barriers to Green Buildings

Still, there are barriers to the widespread adoption of green building techniques, technologies, and systems, some of them related to real-life experience and the rest to perception in the building industry that green buildings still add extra cost. This is surprising because senior executives representing architectural/engineering firms, consultants, developers, building owners, corporate owner-occupants, and educational institutions have held positive attitudes about the benefits and costs of green construction for some time.* For example, I even found a LEED-certified public library in remote Homer, Alaska (Fig. 6.1).

Given these positive views, it is surprising that the leading obstacles to widespread adoption of green buildings continue to be perceived higher costs and lack of awareness of the full range of benefits of green construction and operations. Other factors discouraging green building remain the perceived complexity and cost of LEED documentation; short-term budget horizons on the part of clients and long paybacks for some renewable energy measures; and the often split incentives between commercial building owners and tenants. Sally Wilson of CB Richard Ellis, the world's largest property management company, describes how this might work.†

> I worked on a project where I had a tenant who was going into a LEED-NC building. It was a small tenant taking up about 15,000 square feet. The major tenant was a utility and they had the LEED requirement initially. There were a lot of requirements in the lease that the tenant had to comply with on the LEED side. Helping the tenant understand what the added cost is compared to a non-LEED building was pretty critical. The other important role for us was structuring the lease in a way to get the landlord to take responsibility for payment of those things. The landlord was essentially getting the tax credits but they were trying to get the tenant to pay for the improvements. So that was a negotiation point. The tenant wasn't pursing LEED but they had to understand what it meant to them in terms of the lease and their obligations.

*Turner Construction Company [online], http://www.turnerconstruction.com/greenbuildings/content.asp?d=5785, accessed March 6, 2007.
†Interview with Sally Wilson, February 2008.

Figure 6.1   Designed by ECI/Hyer Architecture & Interiors, with abundant day-lighting and use of local materials, the 17,200-square-feet Homer Public Library in Alaska is LEED-NC Silver certified. Even in one of the smaller towns of Alaska, literally at the end of the paved road from Anchorage, there is a LEED-certified library. It shows that green buildings are viable anywhere. *ECI/Hyer, Inc. and Chris Arend Photography.*

Despite these obstacles, owners and building teams continue to develop LEED Platinum projects, so there must be overriding factors such as strong internal commitments to sustainability.

Paul Meyer is the F. Otto Haas Director of the Morris Arboretum of the University of Pennsylvania. Of his commitment to attaining a LEED Platinum rating for the new arboretum facility, he says:*

> Attaining a LEED Platinum certification is very important to the Arboretum. We want this to really be an exhibit of best sustainable practices. At this point, we're pretty confident that we can attain Platinum. It was something that we were concerned about and planning for since day one. You can't just design a project and then in the end decide that you want it to be LEED certified. LEED goals have to, from the very beginning, be part of the design discussion and process. Early on we did an inventory of [LEED credit] points, so we knew that we could achieve Gold pretty easily. We

---

*Interview with Paul Meyer, February 2008.

also knew it would be a stretch to get to Platinum. At this point [March 2008], we think Platinum is attainable.

Dr. Douglas Treadway of Ohlone College in California described how the overview of the college's mission led to the requirement for a LEED Platinum building.

The goals were determined after series of planning retreats, visioning exercises, interviews and research within the Bay Area to determine the feasibility of certain approaches to green building. We then tied that into the vision the new campus, which also had not been targeted. It was going to be a general college and then we changed it to a health sciences and technology college. We then had a different rationale for our green building, because the nature of the institution's mission had changed. Repurposing the building from a general college campus to a thematic health science and technology campus was really important in the early design because it drove everything after that.

We also consulted with local industry leaders and hospital leaders to get a grasp of where the emerging fields were going and how green architecture would be a part of them. If you go into our new building now and you have asthma, you'll feel instant relief. I just had a person give me testimony to that. As soon as he walked in the door, he realized the air inside was of much better quality than the outside air. He felt an instant relief of some of his symptoms.

OVERCOMING BARRIERS TO GREEN BUILDINGS

Some of the barriers to green building performance have little to do with cost and a lot to do with how engineers and architects have become accustomed to working with each other, much like brothers and sisters in the same family who want nothing to do with each other during adolescence, but who become good friends later in life. Dan Nall is senior vice president of Flack+Kurtz, one of the leading building engineering firms in the United States. Here's his take on the situation.*

Engineers don't really necessarily understand what is that architects are trying to achieve and probably architects, with respect to mechanical engineers, don't understand why we're so concerned about certain things.

The biggest sort of conflict occurs with respect to—and right now this is an issue because of fashion—the amount of glass in the façade of the building and the nature of that glass. Architects don't necessarily understand that making the window wall out of clear glass not only means that the energy consumption of the building is going to go up, but also it makes it almost impossible for the people inside to be comfortable no matter how much air you put on them. They don't understand the basic physics and heat gain occurring from both a conductive standpoint and from a radiance standpoint. That's really an issue. Engineers don't necessarily understand how

---

*Interview with Dan Nall, April 2008.

to talk to architects about the architectural goals that are being sought with this particular type of façade, presenting alternative means of achieving that same visual effect without the compromise of energy efficiency and occupant comfort.

I think being willing to listen and understand the issues is very important. Architects should understand that if they've got a good consultant, the consultant is trying to help them make a good building. The concerns that the good consultant has are legitimate; they're not just tying to be an impediment. I had an architect say to me once a long time ago, "I hate engineers because they're always telling me what I can't do." I thought to myself, "Well if you were able to figure out the things on your own that you shouldn't be doing, then engineers wouldn't have to tell you that."

With a common goal of achieving high-performance LEED Gold and Platinum buildings, architects, engineers, builders, and developers are working hard to bring costs into line with benefits, in three specific ways. Chapter 7 shows the many ways in which design and construction decisions influence the costs of green buildings. Over the next few years, the green building industry is likely to focus on lowering the cost barrier, in several ways:

- Working aggressively to lower the costs of building green through accumulating project experience and strengthening the focus on integrated design approaches that might lower overall costs.
- Marrying capital and operating costs by finding ways to finance green building improvements to reduce or eliminate any first-cost penalty that may occur, using demand reduction incentives from electric utilities, utility "public purpose" programs, and local, state, and federal governments to maximize points of leverage. There are also a growing number of third-party financing sources for onsite generation (combined heat and power or district heating/cooling) and solar power investments in large building projects.
- Review case studies and visit successful projects that have documented the full range of green building benefits so that building owners with a long-term ownership perspective can be motivated to find the additional funds or create the right environment for building high-performance buildings.

Ben Weeks of Aardex describes how they justified the extra costs for underfloor air distribution systems in the Signature Centre project in Denver:

Each design element has many characteristics: first costs, life-cycle costs, training issues. What is the interface with other design elements? What is the time required for installation and what are the other time issues associated with it? Are there competition issues such as how many competitors are available to bid for both materials and installation and, what happens if the selected subcontractor goes broke? Do you have somebody that you replace them with? Is this traditional technology or is it new or cutting-edge technology? Where are the failure opportunities? We consider each of these for every building component in as "holistic" a fashion as possible. Once thoroughly vetted for cost and benefit to the project, we adopt those

Figure 6.2   At the Signature Centre, an underfloor air distribution system supplies air at low pressure and velocity through floor diffusers. The air rises through natural convention and exits through a return air shafts centrally located at the core. This system uses far less energy than conventional overhead air distribution and allows more use of "free cooling" from outside air, because the entering air temperature is typically 8 degrees warmer. *Courtesy of Aardex, LLC.*

technologies and components that make enduring, sustainable contribution and reject those that don't.

For example, we became aware that the benefits of an underfloor air delivery system (Fig. 6.2) were significant and believed it was the way we wanted to go. On the surface, it appeared that the first cost increase was substantial. The cost of the underfloor system alone was about $7 or $8 per square foot, installed. When simply considered for first cost, most developers wouldn't even consider it because it's a huge increase in expense. But in our case, we were able to consider it for many of the things it allowed us to do, such as remove all of the ducting in the building. We essentially have no ducts other than four vertical supply and return ducts. It also allowed us put all of the power and data wiring underneath the floor in a plug-and-play situation, which reduced the amount of electrical material and labor expense substantially. Another major benefit was that it allowed us to reduce the floor-to-floor height by about 10 inches. That 10 inches came out of every floor, times five floors, eliminating 4.2 feet of the total building height. The circumference of the building (1400 square feet) times $76 per square foot for the cost of the skin resulted in a $446,000 savings. The first cost of the underfloor air system was not only reduced, but we spent less on our mechanical systems than a traditional overhead air-distribution system. In addition, the Signature Centre project is located in a zoning district with a height restriction of 75 feet. The height savings afforded by the UFAD (underfloor air delivery) design allowed us to add an extra floor to the building which provides internal and external benefits to all interests.

PLATINUM PROJECT PROFILE

The Armory/Portland Center Stage, Portland, Oregon

Occupied by the Portland Center Stage Theater Company, the Bob and Diana Gerding Theater at the Armory is listed on the National Register of Historic Places. Originally built in 1889 for the Oregon National Guard, the building's 55,000-square-feet, $36 million renovation was completed in September 2006. Chilled beams are the primary cooling system (chilled water comes from a nearby district cooling plant), and high-efficiency gas-fired condensing boilers provide the building's heating needs. The passive chilling and air-circulation features reduce the mechanical system energy use by 40 percent. Rainwater harvesting, no onsite irrigation, dual-flush toilets, and low-flow fixtures have reduced potable water demand by 88 percent. Due to the building's historic façade and existing orientation, neither passive solar design approaches nor photovoltaics were energy-saving options.*

Benefits That Build a Business Case

The business case for green development is based on a framework of benefits: economic, financial, productivity, risk management, public relations and marketing, and funding.† Many people also describe these benefits in terms of the "Triple Bottom Line," with such names as "People, Planet, and Profits." The key issue here is that the benefits will vary by type of ownership, type of use, level of investment and similar drivers. It's very important, in my view, for building team members to become as articulate about the benefit side of the green building equation as they are about the cost

---

*Cascadia Region Green Building Council [online], http://casestudies.cascadiagbc.org/overview.cfm?ProjectID=833, accessed April 2008.

†U.S. Green Building Council, *Making the Business Case for High-Performance Green Buildings* (Washington, D.C.: U.S. Green Building Council, 2002), available at: www.usgbc.org/resources/usbgc_brochures.asp[0], accessed March 6, 2007. See also *Environmental Building News*, 14, no. 4 (April 2005), available at: www.buildinggreen.com, accessed March 6, 2007.

side, to become as knowledgeable about financial benefits and considerations as they are about the technical issues concerning their own specialties. I have found professionals in all disciplines that are unable to clearly articulate the business side of green buildings. In my view, a critical professional skill is to learn how your clients make money and to become fluent in the language of business and investment. Most clients take it for granted that you'll be able to do a good job of design and construction. However, they want you to understand and articulate the reasons for green building in terms that make sense to them and which they can report to key people in their own organization.

Table 6.1 presents a sample of the wide-ranging benefits of green buildings, examined in detail in the following sections.

TABLE 6.1   BENEFITS OF GREEN BUILDINGS

1. Utility cost savings for energy and water, typically 30 to 50 percent, along with reduced "carbon footprint" from energy savings

2. Maintenance cost reductions from commissioning, operator training, and other measures to improve and ensure proper systems integration and ongoing performance monitoring

3. Increased value from higher net operating income (NOI) and increased public relations for commercial buildings

4. Tax benefits for specific green building investments such as energy conservation and solar power, and local incentives, depending on location

5. More competitive real estate holdings for private sector owners, over the long run, including higher resale value (see the CoStar study cited earlier)

6. Productivity improvements for long-term building owners, typically 3 to 5 percent

7. Health benefits, including reduced absenteeism, typically 5 percent or more

8. Risk management, including faster lease up and sales for private developers, and less risk of employee exposure to irritating or toxic chemicals in building materials, furniture, and furnishings

9. Marketing benefits, especially for developers, large corporations, and consumer products companies

10. Public relations benefits, especially for developers and public agencies

11. Recruitment and retention of key employees and higher morale

12. Fund raising for colleges and nonprofits

13. Increased availability of both debt and equity funding for developers

14. Demonstration of commitment to sustainability and environmental stewardship; shared values with key stakeholders

---

PLATINUM PROJECT PROFILE

Inland Empire Utilities Agency, Chino, California

The Inland Empire Utilities Agency headquarters in Chino, California was a two building project with a total of 66,000 square feet. Total project cost was $7.5 million. The construction cost for the two buildings at $154 per square foot, was far below the industry standard of $180 to $294 for comparable buildings at that time. The agency expects to save $800,000 in energy savings annually. A roof-mounted photovoltaic system is expected to produce more than 100,000 kilowatt-hours of electricity per year. Both buildings are heated and cooled by heat recovered from power generators at the neighboring water-recycling plant. Reclaimed water from the plant as well as building and stormwater harvesting is reused for toilets and irrigation reducing potable water consumption by 73 percent (compared to a conventional building).*

Economic Benefits

Increased economic benefits are the prime driver of change for green buildings. In fact, relative economic advantage is the prime driver of almost all innovation. Economic benefits of green buildings vary considerably, depending on the nature of the building ownership, and their full consideration is vital for promoting any sustainable design.

REDUCED OPERATING COSTS

With the real price of oil likely to stay above $80 to $100 per barrel (in today's dollars) for the forseeable future,[†] natural gas prices at near-record levels, and peak-period (typically summer air-conditioning times) electricity prices rising steadily in many metropolitan areas, energy-efficient buildings make good business sense. Even in commercial "triple-net" leases (the most common type) in which the tenant pays all operating costs, landlords want to offer tenants the most economical space for their money (not doing so amounts to a hidden tax on the renter). For small added investments in capital cost, green buildings offer lower operating costs for years to come. Many green buildings are designed to use 30 to 50 percent less energy than current codes require; some buildings achieve even higher efficiency levels. Translated to an operating cost of $3.00 per square foot for electricity (the most common source of energy for office and commercial buildings), this level of savings could reduce utility

---

*U.S. Green Building Council [online], http://leedcasestudies.usgbc.org/overview.cfm?ProjectID=278, accessed April 2008.
†U.S. Energy Information Administration [online]. For the November 2006 forecast, see www.eia.doe.gov/oiaf/aeo/key.html, accessed March 6, 2007.

operating costs by $0.90 to $1.50 per square foot per year. Often these savings are achieved for an investment of just $1.00 to $3.00 per square foot. With building costs reaching $150 to $300 per square foot, many developers, institutions, and building owners consider it a wise business decision to invest 1 to 2 percent of capital cost to secure long-term savings, particularly with a payback of less than 3 years. In an 80,000-square-foot building, this level of owner's savings translates into $72,000 to $120,000 per year, year after year.

REDUCED MAINTENANCE COSTS

More than 120 studies have documented that a properly commissioned building shows additional energy cost savings of 10 percent to 15 percent. These buildings also tend to be much easier to operate and maintain.* By conducting comprehensive functional testing of all energy-using systems before occupancy, it's often possible to have a smoother-running building for years because potential problems are fixed in advance. A recent review of these studies by Lawrence Berkeley National Laboratory showed that the payback from building commissioning in terms of energy savings alone was about 4 years, while the payback fell to about 1 year when other benefits were considered, such as fewer callbacks to address thermal comfort complaints.

INCREASED BUILDING VALUE

Increased annual energy savings also create higher building values. Imagine a building that saves $72,000 per year in energy costs versus a comparable building built to code (this savings might result from saving only $0.90 per year per square foot for an 80,000-square-foot building). At capitalization rates of 6 percent, typical today in commercial real estate, green building upgrades would add $1,200,000 ($15 per square foot) to the value of the building. For a small upfront investment, an owner can reap benefits that typically offer a payback of three years or less and an internal rate of return exceeding 20 percent. The CoStar study cited earlier and others show that ENERGY STAR-rated and LEED-certified buildings exhibit resale premiums of up to 30 percent.

TAX BENEFITS AND INCENTIVES

More than 600 incentive programs for renewable energy and energy efficiency are offered by all levels of government.† Some states offer tax benefits for green buildings. Oregon and New York offer state tax credits. New York's tax credit allows builders who meet energy goals and use environmentally preferable materials to claim credits

---

*Lawrence Berkeley National Laboratory, *The Cost-Effectiveness of Commercial-Buildings Commissioning*, 2004 [online], http://eetd.lbl.gov/emills/PUBS/Cx-Costs-Benefits.html, accessed April 2008. This research reviewed 224 studies of the benefits of building commissioning and concluded that based on energy savings alone, such investments have a payback within 5 years.

†The best source is the database of State Incentives, www.dsireusa.org, accessed June 30, 2008.

up to $3.75 per square foot for interior work and $7.50 per square foot for exterior work against their state tax bill. To qualify for the credit, in new buildings, energy use cannot exceed 65 percent of use permitted under the New York State energy code; in rehabilitated buildings, energy use cannot exceed 75 percent.*

The state of Nevada offers a property tax abatement of up to 35 percent reduction, for up to 10 years, for private development projects achieving a LEED Silver certification. Assuming the property tax is assessed at 1 percent of value, this abatement could be worth as much as 3.5 percent of the building cost, typically far more than the actual cost of achieving LEED Silver on a large project. As a result, a large number of Nevada projects are pursuing LEED certification.[†] The Nevada law also provides for sales tax abatement for green materials used in LEED Silver–certified buildings.

The 2005 federal Energy Policy Act offers two major tax incentives for aspects of green buildings: a tax credit of 30 percent on both solar thermal and electric systems and a tax deduction of up to $1.80 per square foot for projects that reduce energy use for lighting, HVAC and water heating systems by 50 percent compared with a baseline standard.[‡] This law is scheduled to expire at the end of 2008, and as of the spring of 2008, the deadline had not yet been extended by the Congress, although many expect it to occur.

PRODUCTIVITY GAINS

In the service economy, productivity gains for healthier indoor spaces are worth anywhere from 1 to 5 percent of employee costs, or about $3.00 to $30.00 per square foot of leasable or usable space. This estimate uses average employee costs of $300 to $600 per square foot per year (based on $60,000 average annual salary and benefits and 100 square feet to 200 square feet per person).[§] With energy costs typically less than $3.00 per square foot per year, it appears that productivity gains from green buildings could easily equal or exceed the entire energy cost of operating a building.

Figure 6.3 shows that median productivity gains from high-performance lighting are 3.2 percent in 11 studies analyzed by Carnegie-Mellon University in Pittsburgh, or about $1 to $2 per square foot per year, an amount nearly equal to the cost of energy.[¶] This benefit is in addition to a reported average savings of 18 percent on total energy bills from proper lighting. For corporate and institutional owners and occupiers of buildings, that is too much benefit to ignore during the design process.

---

*Natural Resources Defense Council [online] www.nrdc.org/cities/building/nnytax.asp[0], accessed March 6, 2007.

[†]Personal communication, Lynn Simon, Simon & Associates, February 2, 2007. Also see US Department of Energy [online], www.eere.energy.gov/states/news_detail.cfm/news_id=9149, accessed March 6, 2007 and http://www.leg.state.nv.us/22ndSpecial/bills/AB/AB3_EN.pdf, accessed March 6, 2007.

[‡]U.S. Department of Energy [online], www.energy.gov/taxbreaks.htm, accessed March 6, 2007.

[§]Eleven case studies have shown that innovative daylighting systems can pay for themselves in less than 1 year due to energy and productivity benefits. Vivian Loftness et al., Building Investment Decision Support (BIDS) (Pittsburgh: Center for Building Performance and Diagnostics, Carnegie Mellon University, n.d.), available at http://cbpd.arc.cmu.edu/ebids, accessed March 6, 2007.

[¶]Carnegie Mellon University, http://cbpd.arc.cmu.edu/ebids/images/group/cases/lighting.pdf, accessed March 6, 2007.

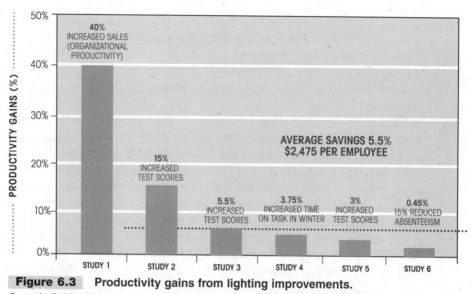

Figure 6.3   Productivity gains from lighting improvements.

Center for Building Performance and Diagnostics, Carnegie Mellon University. EBIDS: Energy Building Investment Decision Support Tool.

Look at it this way: If a building owner could get a 10 percent improvement in productivity from a green building, or about $30 to $60 per square foot increase in annual output, it would almost always pay for that company to build a new building and have employees to work there. In other words, the productivity increase could pay for the entire building! Even a 5 percent improvement in productivity would pay for half or more of the rent or cost of the new green building. What, then, is the business case for a "brown building," a standard building without these benefits?

In 2003, a detailed study of 33 LEED-certified green buildings provided a 20-year "net present value" calculation for the various categories of green building benefits.* In that study, productivity and health gains provided more than two-thirds of the total benefits of green buildings in this analysis. Energy and maintenance savings provided an additional 20 percent to 25 percent of total benefits.† The concept of using net present value (the discounted value of all project cash flows) has greatest relevance for long-term owner-occupiers of buildings, such as government agencies, large corporations, universities, schools, and nonprofits that are likely to enjoy the full benefits over time.

Risk Management

Consider the possibility that green building certification could provide some measure of protection against future lawsuits through third-party verification of measures installed to protect indoor air quality, beyond just meeting code-required minimums.

---

*Capital E Consultants [online], www.cap-e.com/ewebeditpro/items/O59F3303.ppt#2, accessed March 6, 2007.
†Gregory Kats, "Financial Costs and Benefits of Green Buildings," [online] www.cap-e.com.

(This section is not intended to give you legal advice—please consult with your own attorney regarding risk mitigation issues!) With a national focus on mold and its effect on building occupants, developers and building owners should focus considerable attention on improving and maintaining indoor air quality. If green buildings could give you a better risk management posture, what would that be worth? Five percent of initial cost, 1 percent?

The ability of green buildings to qualify for faster permitting or special permit assistance in a number of cities can also be considered a type of risk mitigation. In Chicago, for example, the city government has created the position of green projects administrator and offers green projects priority processing of building permits. For large projects, above minimum requirements, the city waives fees for independent code consultants. Projects with high-level green goals are promised a 15-day permit review.[*] In Austin, Texas, the city fast-tracked the development reviews for a large big box retailer, so that it was able to open 12 months ahead of schedule; the resulting profit gain of about $3 million reportedly paid for the entire $2.8 million building![†]

In 2007, I conducted a study for NAIOP, a national property developer's organization, and found dozens of jurisdictions offering green building incentives. That number continues to increase each year.[‡] Local governments offered incentives ranging from accelerated land use and building permit review to density bonuses and tax abatements. (These benefits typically don't apply to governmental and institutional projects which may not require such local permits.)

Another risk management benefit of green buildings in the private sector is the potential for faster sales and leasing of such buildings, compared to similar projects in the same city. Green buildings tend to be easier to rent and sell, because educated tenants increasingly understand the benefits (see the CoStar study cited above). Many corporate real estate departments are now beginning to require LEED certification wherever possible, as a condition of leasing space in commercial buildings. In some cases, a building may be fully leased before construction completion, reducing the developer's "market risk." Imagine the benefit to a developer from having all leases signed, before having to incur all the costs of construction.

Green buildings are also seen as less risky by some insurers. In September of 2006, Fireman's Fund, a major insurance company, announced it would give a 5 percent reduction in insurance premiums for green buildings. The insurer also announced its Certified Green Building Replacement and Green Upgrade coverage.[§] After this initial announcement, other commercial insurance carriers also began offering special insurance benefits to certified green buildings. While not representing huge savings, these premium reductions add to the benefit side of the green building ledger.

---

[*] "Speedy Permitting Has Developers Turning Green in Chicago," *Building Design & Construction*, November 2005, p. 28; www.BDCnetwork.com, accessed March 6, 2007.
[†] Personal communication, S. Richard Fedrizzi, CEO, U.S. Green Building Council.
[‡] NAIOP, www.naiop.org/foundation/greenincentives.pdf.
[§] www.buildingonline.com/news/viewnews.pl?id=5514, accessed March 6, 2007.

PLATINUM PROJECT PROFILE

Laurence S. Rockefeller Preserve, Grand Teton National Park, Wyoming

The Laurence S. Rockefeller Preserve in Grand Teton serves as a visitor center. Located near Jackson Hole, Wyoming, the 7500-square-feet building houses an interpretive center for the National Park Service. The building's design should reduce energy use by 84 percent and save $2000 annually. A ground-source heat pump and photovoltaic system is expected to provide 58 percent of the center's electricity. All of the wood used in construction was certified to FSC standards. The restrooms at the facility use composting toilets, saving an estimated 76,000 gallons of water annually.*

Health Improvements

Of course, a key element of productivity is healthy workers. By focusing on measures to improve indoor environmental quality such as increased ventilation, daylighting, views to the outdoors, and low-toxicity finishes and furniture, people in green buildings show an average reduction in symptoms of 41.5 percent on an annual basis.[†] Obviously, healthy employees are more productive, since they're at work more and likely to be performing at a higher energy level than those who are sick or not feeling well.

Since most companies are effectively self-insured (i.e., your health insurance costs go up the more claims you have) and most government agencies and large companies are actually self-insured, it makes good economic sense to be concerned about the effect of building design on people's health. In addition, given what is already known about the health effects of various green building measures, a company might be inviting lawsuits if it didn't take all feasible measures to design and construct a healthy building.

Public Relations and Marketing

Many developers and building owners, both public sector and private companies, are finding considerable marketing and public relations benefits from creating LEED-certified green buildings. National media in the United States and Canada have effectively bought the proposition that LEED-certified buildings represent better buildings and a stronger commitment to sustainability by building owners and developers. LEED

---

*World Clean Energy Awards [online], http://www.cleanenergyawards.com/top-navigation/nominees-projects/nominee-detail/project/41/, accessed April 2008. Building Green [online], http://www.buildinggreen.com/auth/article.cfm/2008/1/1/Grand-Teton-Visitor-Center-Earns-LEED-Platinum/?, accessed April 2008.
[†]Center for Building Performance and Diagnostics, Carnegie Mellon University. eBIDS: Energy Building Investment Decision Support Tool, http://cbpd.arc.cmu.edu/ebids, accessed April 30, 2008.

certification also provides effective protection against charges of "greenwashing," or making claims of environmental benefits that are exaggerated or can't be substantiated.

Because positive public relation has direct monetary benefits for private developers, in terms of gaining government approvals or mitigating citizen opposition to projects, it is a critical benefit of green buildings for them. In addition, many institutional owners (hospitals, universities, K-12 schools, and similar facilities) are vitally dependent on public relations for fund-raising, legislative budget authorization, bond elections, and other purposes; it is in their interest to put forward a green building agenda as a sign of an organizational commitment to sustainability.

STAKEHOLDER RELATIONS AND OCCUPANT SATISFACTION

Those who work in buildings as tenants and corporate, public and institutional employees want to see a demonstrated concern for their well-being and for that of the planet. Intelligent owners are beginning to realize how to market these benefits to a discerning and skeptical client and stakeholder base, using the advantages of green building certifications and other forms of documentation, including support from local utility and industry programs. This is more than just "greenwashing," it is a positive response to a growing public concern for the long-term health of the environment. A good indication of how corporations have embraced this concept is the explosion in green building projects and associated public relations since 2006; for example, if you sign up to receive Google Alerts and put in "green buildings" as a keyword, you will be inundated with 6 to 12 news stories almost every day from the news media, as well as numerous blog entries.

PLATINUM PROJECT PROFILE

Bernheim Arboretum Visitors Center, Clermont, Kentucky

"Imagine a building like a tree" was the design concept for the Bernheim Arboretum Visitors Center, completed in 2005. This 6000-square-feet facility contains recycled wood (from pickle vats and bourbon whiskey rack houses) that may be the building's most apparent sustainable feature. Simulating the natural processes of the tree, the facility has a peat-moss sewage filtration system and an 8000-gallon underground cistern. Wrap-around trellises and pergolas provide shade and, like a tree, natural habitat for local flora and fauna. The design employs a green roof and both active and passive solar systems.*

ENVIRONMENTAL STEWARDSHIP

Being a good neighbor is appropriate not just for building users, but for the larger community. Developers, large corporations, universities, schools, local governments,

---

*Zach Mortice, "Arboretum Visitor Center Stands Tall—Against the Yardstick of a Tree", AIArchitect, Volume 15, February 15, 2008 [online], http://www.aia.org/aiarchitect/thisweek08/0215/0215d_bernheim.cfm, accessed April 2008.

and building owners and managers have long recognized the marketing and public relations benefits (including branding) of a demonstrated concern for the environment. Green buildings fit right in with this message. As a result, I expect to see a growing number of major commitments by corporate real estate executives to green their buildings and facilities. A good example is Adobe Systems, Inc., a major software maker based in San Jose, California. In 2006, Adobe announced that it had received three LEED-EB Platinum awards for its headquarters towers; not only did it reap great publicity, but the firm showed that the investments as a whole had benefits with a net present value almost 20 times their initial cost.*

Many larger public and private organizations have well-articulated sustainability mission statements and are understanding how their real estate choices can both reflect and advance those missions. For example, in 2007 the world's largest property developer CB Richard Ellis committed to certify more than 100 properties under the LEED-EB standard. Developer Jonathan F. P. Rose notes that "having a socially and environmentally motivated mission makes it easier for businesses in the real estate industry to recruit and retain top talent. Communities are more likely to support green projects than traditional projects, and it is easier for such projects to qualify for many government contracts, subsidies, grants, and tax credits. The real estate industry can prosper by making environmentally responsible decisions."†

Green buildings also reinforce a company's brand image. A consumer retailer such as Wal-Mart, Kohl's, Office Depot, Best Buy, Starbucks, or PNC Bank can improve or maintain their brand image by being associated with green buildings, and so they are moving in this direction.‡ Large corporations, including those that issue sustainability reports every year or two—in 2007, there were more than 1200 of them according to the Global Reporting Initiative—see the benefits of building green to demonstrate to their employees, shareholders, and other stakeholders that they are "walking the talk."

MORE COMPETITIVE PRODUCT

Speculative developers are realizing that green buildings can be more competitive in certain markets, if built at or near a conventional budget. This puts the onus on the design and construction team to figure out how to build LEED Silver or better buildings without adding significant costs. Green buildings with lower operating costs and better indoor environmental quality are more attractive to a growing group of corporate, public, and individual buyers. *Green-ness* will not soon replace known real-estate attributes such as price, location, and conventional amenities, but green features will increasingly enter into tenants' decisions about leasing space and into buyers' decisions about purchasing properties and homes. It's noteworthy that the largest commercial real

---

*U.S. Green Building Council [online] www.usgbc.org/News/PressReleaseDetails.aspx?ID=2783, accessed March 6, 2007.

†"The Business Case for Green Building," *Urban Land*, June 2005, p. 71; www.uli.org, accessed April 30, 2008.

‡For example, PNC Bank has committed to making all of its new branches LEED-certified, at least at the basic level.

estate database, CoStar, which has information on more than two million properties, since 2006 has listed whether a given office property has a LEED certification or ENERGY STAR rating.

A study of 2000 large office buildings (defined as having at least 200,000 square feet of leasable area) in the CoStar database by Professor Norman Miller of the University of San Diego, released late in 2007, showed that over the period of 2004 through the first half of 2007, ENERGY STAR buildings (in the top 25 percent of energy efficiency in the United States) had $2.00 per square foot higher rents and 2 percent higher occupancy. Each of these two factors increases building value measurably. In fact, the ENERGY STAR-rated buildings that were sold in 2006 commanded a 30 percent premium (sales price per square foot) over those buildings that weren't so rated.* This is consistent with the larger CoStar study cited at the beginning of this chapter.

There are developers using the precertification available from the LEED for Core and Shell rating system to attract tenants and financing for high-rise office towers. A large office tower in Chicago that opened in 2005 received a LEED-CS Gold rating along with the marketplace benefits of fast leasing and great tenants. Designed by Goettsch Partners for the John Buck Company, the 51-story tower contains 1,456,000 square feet (134,000 square meter) of space, including a 370-car parking garage. The building, *111 South Wacker*, is anchored by the professional services firm Deloitte, which leased 451,259 square feet (41,440 square meter), or more than 43 percent of the building.

The marketing benefits of LEED-CS pre-certification have encouraged nearly 1000 projects to register for this program by the spring of 2008; since these buildings average about $50 million in value (typically more than 350,000 square feet in area), this is a strong statement from the development community about the expected value of LEED certification.

Recruitment and Retention

One often overlooked aspect of green buildings is their effect on people's willingness to join or stay with an organization. It typically costs $50,000 to $150,000 to lose a good employee, and most organizations experience 10 to 20 percent turnover per year, some of it from people they really didn't want to lose. In some cases, people leave because of poor physical environments (as well as the Dilbert-parodied "boss from hell"). In a workforce of 200 people, turnover at this level implies 20 to 40 people leaving per year.

What if a green building could reduce turnover by 5 percent, for example, one to two people out of the 20 to 40? Taken alone, that value would range from $50,000 to possibly as much as $300,000, more than enough to justify the costs of certifying a

---

*See, for example, Daily Commercial News, http://dcnonl.com/article/id26427, accessed June 30, 2008.

building project. If a professional service firm, say a law firm, retained just one good attorney, typically billing $300,000 to $400,000 per year, with $200,000 to $250,000 gross profit, that would more than pay for the extra cost of a green building or green tenant improvement project, if such measures would keep that lawyer at the firm. What about the impact of a healthy work environment on employees' belief that their employer really cares about their well-being? One study of 2000 office workers, commissioned in 2006 by the architecture firm Gensler, revealed that nearly half of all respondents would be embarrassed to show their offices to clients or prospective employees. What does this say about morale and their feeling that employer really cares about their well-being?

Demographics is destiny: Owing to an aging Baby Boomer labor force, by 2014 there will be fewer people in the 35 to 44-year-old age group than in 2005, typically the leadership group in most organizations: managers, executives, experienced employees, and senior technical people, typically at the peak of their career. Getting and keeping them will tax the ingenuity and resources of most companies; green buildings can demonstrate that the company or organization and the key employees share the same values. Working in a company that rents or owns green buildings gives employees another reason to tell their friends and spouses why they are staying with an organization.

PLATINUM PROJECT PROFILE

Genzyme Center, Cambridge, Massachusetts

Designed in the early 2000s by Behnisch Architekten of Stuttgart, Germany and finished in 2004, the 12-story, 344,000-square-feet Genzyme Center serves as office space for more than 900 Genzyme employees. Located in Cambridge's Kendall Square, the building is partially powered by renewable energy sources and controlled by a $2.3 million integrated building automation system (BAS). An open-air atrium serves a return air duct as well as light shaft and provides natural light to 75 percent of the employee workspaces, which contributes to a 42 percent electricity cost reduction. Fully automated, roof-mounted heliostats (mirrors) track the sun's movement and shine light into the atrium. Fixed mirrors, hanging prismatic mobiles, reflective panels, and a reflective light wall all work together through an automated computerized system to reflect and diffuse natural light throughout the building. Water-free urinals, dual-flush toilets, automatic faucets, low-flow fixtures, and stormwater-supplemented cooling tower water supply collectively reduce water use by 32 percent. Ninety percent of the wood is FSC certified.*

---

*Evelyn Lee, "Green Building: Genzyme Center LEEDs the Way", Inhabitat, February 6, 2007 [online], http://www.inhabitat.com/2007/02/06/genzyme-center/, accessed April 2008. Building Green [online], http://www.buildinggreen.com/hpb/overview.cfm?projectId=274, accessed April 2008.

Photography by Stefan Behnisch.

Financing Green Projects

Whether public or private, raising money for new and renovated projects is always an issue. For private developers, raising both debt and equity capital is typically the challenge. The rise of socially responsible property investing promises to reward those developers who build green.

Investing in green buildings has begun to attract considerable attention as a form of responsible property investing (RPI), a practice which is growing faster than overall investing. One expert, Professor Gary Pivo, puts it this way in a 2007 report on a survey of 189 leading real estate executives in the United States:*

There is considerable interest and activity in RPI in America's real estate investment organizations. It reflects the commitment to sustainability and corporate social responsibility that appears to be taking hold in the business world at large. As CoreNet

---

*Gary Pivo, "Responsible Property Investing: A Survey of American Executives," www.u.arizona.edu/~gpivo/ RPI%20Survey%20Brief.pdf, accessed April 23, 2008.

Global put it in the introduction to their recent conference on sustainability and real estate, "The trend toward taking the Triple Bottom Line approach to business continues to accelerate, and we have reached a tipping point—sustainability of people, planet, and profit is now a mandate for multinational companies operating in a global economy. What does this mean for practitioners of real estate and workplace strategy?"

We are beginning to see what it means in these results. Most property investment executives say they're going beyond minimum legal requirements to address social or environmental issues. Many are promoting energy and natural resource conservation, engaging with key stakeholders affected by their work, and recognizing sustainability and social responsibility in business strategies and value statements. A third or more say they've invested in socially and environmentally beneficial properties like urban infill, green buildings, brownfields, and transit-oriented development. More than a third say their organization recognizes the efficiencies associated with RPI. Another 30 percent go farther, saying it's in their self-interest to make it part of their business strategy. And 10 percent report being Sustaining Organizations that are fundamentally committed to RPI and actively promoting it in business and society.

What's driving this apparent transformation? According to our real estate's top executives the primary drivers are business concerns: avoiding risks associated with environmental or social problems that could harm returns and seeking opportunities associated with consumer interest in health, community, equity, and ecology. Although ethics and volunteerism are also at play, the importance of business motivation in the process bodes well for the future of RPI.

In 2006, New York–based developer Jonathan Rose created the *Rose Smart Growth Investment Fund* to invest in green building projects. The $100 million limited partnership focuses on acquiring existing properties near mass transit. The fund expects to make green improvements to the properties and hold them as long-term investments.* The focus on transit-centric developments takes into account the energy savings from enabling people to use mass transit. The fund's first project is in downtown Seattle, Washington, a renovation of the 1920s-era Joseph Vance and Sterling buildings, a total building area of about 120,000 square feet, with ground-floor retail and office space above.† According to the Fund, it is "rebranding these buildings as the 'greenest and healthiest' historic buildings in the marketplace, to increase market awareness of the buildings, attract and retain tenants."

Many nonprofits have successfully used greening their buildings to attract funds for renovation projects. In 2000, the Ecotrust nonprofit in Portland, Oregon, received a major gift from a single donor to renovate a 100-year-old, two-story brick warehouse into a three-story, 70,000-square-foot modern building with two floors of offices above ground-floor retail. The Jean Vollum Natural Capital Center was only the second LEED Gold–certified project in the United States when it opened in 2001.‡ LEED buildings are being built by a wide range of nonprofit groups, including the Chicago Center for Neighborhood Technology; the Boston-based Artists for Humanity; the Natural

---

New York Times, January 10, 2007.
†Jonathan Rose Companies [online], www.rose-network.com/projects/index.html, accessed March 6, 2007.
‡Ecotrust [online], www.ecotrust.org/ncc/index.html, accessed March 6, 2007.

Resources Defense Council in Santa Monica, California; the Aldo Leopold Legacy Center in Baraboo, Wisconsin; Heifer International in Little Rock, Arkansas; the National Audubon Society in Los Angeles, California; the Chesapeake Bay Foundation in Annapolis, Maryland; and the William A Kerr Foundation in St. Louis, Missouri.

PLATINUM PROJECT PROFILE

The Aldo Leopold Legacy Center, Baraboo, Wisconsin

In 2007, the Aldo Leopold Center became the highest-scoring LEED-NC Platinum project (with 61 out of 69 possible points). Designed by Kubala Washatko Architects, this project serves as the Aldo Leopold Foundation's headquarters, with office and meeting spaces, an interpretive exhibit hall, archive, workshop, and three-season hall. The 12,000 square-foot complex cost $4 million. The facility uses 70 percent less energy than a standard building. A 198-panel, 39.6-kW photovoltaic system was designed to meet 110 percent of the building's annual energy needs. Over 90,000 board-feet of site-harvested wood was used for structural timbers, doors, windows, finish materials, and artisan-crafted furniture. An underground earth-tube system separates ventilation from the heating and cooling systems, savings two to five times the energy of a combined system and allowing the building to use 100 percent fresh air ventilation, even in a rather severe winter climate.*

©The Kubala Washatko Architects, Inc. / Mark F. Heffron.

*www.architectureweek.com/2007/1114/design_4-1.html, accessed April 23, 2008.

Political

Political benefits might be seen as a subset of marketing and public relations benefits, but I'd like green builders to consider them as separate, both for a local political body, as well as for developers relying on public approval for acquiring development entitlements and even design approval and building permits for specific projects. On Earth Day 2008, I saw a spate of political announcements taking place at green buildings, where mayors and various public officials took advantage of the day to pronounce their support for green buildings. In Los Angeles, for example, Mayor Antonio Villaraigosa announced the LEED Gold certification of the Luma condominiums in the downtown area.* The project expects to save 30 percent of energy compared with a traditional building and is part of a major new high-rise housing development just south of downtown Los Angeles, for which all the buildings will be LEED Gold certified.

Who Benefits?

One of the biggest issues in green buildings is that the benefits are unequally distributed between those who pay for the project and those who benefit. For example, the benefits of green schools accrue most directly to the students, but it's the school district (and the taxpayers) who incur the cost. One can argue that the school district should look favorably toward green buildings that benefit students and teachers primarily, but that's not always the case—districts are just as concerned about initial costs and the effect of green buildings on their stretched capital budgets as are developers.

In speculative commercial development, the tenants receive most of the benefits of reduced operating costs and higher productivity, but the developer must bear the initial cost increase. The recent studies cited earlier in this chapter show greater occupancy and higher rents in commercial ENERGY STAR–certified office buildings, as well as higher resale values are compelling, but many developers are concerned whether these benefits will accrue to them. In retail development, for example, shopping center developers in 2008 still do not expect higher rents for LEED-certified retail projects, even though they might enjoy faster political approval and greater marketing and public relations benefits.[†] Nonetheless, major retail developers are announcing LEED project commitments with regularity in 2008, leading one to presume that they see commercial benefits in such activity.[‡]

Table 6.2 shows the distribution of green building benefits; when promoting green buildings to various stakeholders, you should always consider these distinctions in presenting the case for green buildings. Public policy for green buildings should take

---

*http://mayor.lacity.org/villaraigosaplan/EnergyandEnvironment/Greenbuilding/index.htm, accessed April 2008.
[†]Personal communication, Scott Wilson, vice president, construction, Regency Centers, September 2007.
[‡]Vestar Earth Day 2008 announcement; Regency Centers November 1, 2007 announcement.

TABLE 6.2 DISTRIBUTION OF GREEN BUILDING BENEFITS

OWNER TYPE	ENERGY SAVINGS	PRODUCTIVITY GAINS	HEALTH BENEFITS	MARKETING/ PR	RECRUITMENT	FINANCING
Private, owner occupied	Yes	Yes	Yes	Yes	Yes	Yes
Private, speculative office	No	No	No	Yes	No	Yes
Retail, Big box	Yes	No	Maybe	Yes	No	Maybe
Retail, Shopping center	Small percentage	No	No	Yes	Yes	Maybe
K-12 public school	Yes	Yes, for staff	Yes, for staff	Yes	Yes, for teachers	No
Private college	Yes	Yes	Yes	Yes	Yes, for faculty	Possibly, new donors
Public university	Yes	Yes	Yes	Yes	Yes	Yes, for private donors
Nonprofit healthcare	Yes	Yes, for staff	Yes	Yes	Yes, for nurses	Typically not
Nonprofit organization	Yes	Yes	Yes	Yes, very important	Not too important	Yes, for donors
Federal government	Yes	Yes	Yes	Not too important	Not too important	No
State government	Yes	Yes	Yes	Not too important	Not too important	No
Local government	Yes	Yes	Yes	Somewhat important	Somewhat important	Not very important

the distribution of benefits into account and create incentives to overcome gaps in the marketplace. For example, faster permit processing for speculative development can have a huge impact on project returns and is generally a strong incentive that costs the government relatively little.

PLATINUM PROJECT PROFILE

Heifer International, Little Rock, Arkansas

Heifer's mission is to end hunger and poverty while caring for the earth. The organization's 94,000-square-feet, five-story headquarters building was completed in February 2006 at a cost of $17.4 million. Heifer expects to reduce energy consumption by 40 percent (compared to a conventional building). Only 62 feet wide, the orientation and curved shape allow natural light to penetrate the building, reducing the need for artificial light. Rainwater from the permeable parking lot is collected in constructed wetlands, which store and purify up to four million gallons of water for later use. A 25,000-gallon water tower captures rainwater from the roof for use in toilets, the radiant heating system and replenishing the constructed wetland. Recycled, renewable, and regional materials were used throughout the building.*

©Photography by Timothy Hursley.

---

*Heifer International [online], http://www.heifer.org/site/c.edJRKQNiFiG/b.1484715, accessed April 2008.

7

COSTS OF GREEN BUILDINGS

The key benefit of integrated design process is its ability to achieve higher-performance results without significantly increasing overall building costs. Costs are the single most important factor in the development and construction world. The reason is simple: Design and construction costs are "hard" because they are real and occur in the present, whereas benefits such as projected energy savings, water savings, and productivity gains are "soft" because they are speculative and occur in the future. Therefore, a benefit-cost analysis for each project is crucially important, to convince building owners, design teams, and developers to proceed with sustainable design measures and the LEED certification effort. This approach is addressed more fully in Chap. 8.

The biggest barrier to green buildings is the *perception* that they cost more. A survey in the summer of 2007 revealed that 89 percent of respondents, comprising experienced executives and participants in the building design and construction industry, believed that green buildings cost more, with 41 percent believing that they cost more than 10 percent additional.* The World Business Council for Sustainable Development reported similar results in an international survey in the summer of 2007. Respondents to a 1400-person global survey estimated the additional cost of building green at 17 percent above conventional construction! At the same time, survey respondents put greenhouse gas emissions from buildings at 19 percent of world total, while the actual number of 40 percent is double this, counting both residential and commercial buildings.[†]

These surveys reveal that even experienced construction industry participants have false perceptions about a business they know so well. Therefore, the only thing that will overcome this perception of higher cost is to demonstrate that the integrated design process can deliver high-performance buildings at conventional costs. Many of the interviews we conducted for this book verified that people who "know what they

---

Building Design & Construction, 2007 Green Building White Paper, page 8, available at www.bdcnetwork.com, accessed April 22, 2008.

[†]World Business Council for Sustainable Development, "Energy Efficiency in Buildings: Business Realities and Opportunities," August 2007, available at www.wbcsd.org/plugins/DocSearch/details.asp?type=DocDet&ObjectId= MjU5MTE, accessed April 22, 2008.

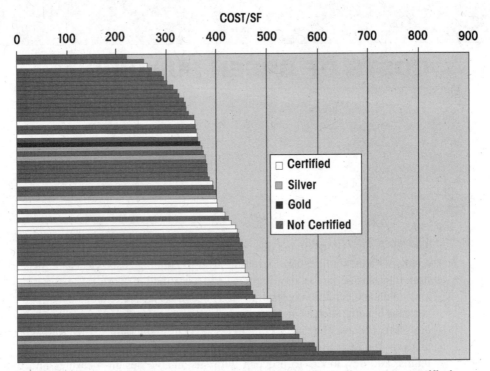

Figure 7.1   According to Davis Langdon's research studies, green-certified labs don't have to cost more than conventional labs. *Courtesy of Davis Langdon.*

are doing" can in fact get these results, no matter what process they follow. What's needed is a substantial process change that works for the balance of building teams.

Lisa Matthiessen is an architect and senior consultant with Davis Langdon, an international cost management firm. Together with her colleagues at Davis Langdon, she has prepared several dramatic costs studies of green buildings. Figure 7.1 shows the results of one of those studies, in this case analyzing the costs of laboratory buildings. The essential message is that you can't tell the difference, in terms of cost per square foot, between LEED-certified and non-certified buildings. This is an interesting study, because it shows that "green" is a program issue, that is, if it's an essential goal of a project, there need be no cost premium. By contrast, when green is treated as an add-on or afterthought, it's likely going to cost more.

Matthiessen's observation is that there is a strong link between the sophistication of the design team, their knowledge of and commitment to integrated design, and the eventual project cost.*

The cost to do LEED projects is comparable to non-LEED projects. It might cost more sometimes, but it's comparable and you usually end up making a trade-off.

*Interview with Lisa Matthiessen, March 2008.

We've seen a corollary between better cost control—lower costs on projects where sustainability really was integrated in—for example, in projects where the team has been able to take simpler moves to get to sustainability, which in turn translates to less money because it usually means fewer or simpler systems. The assumption that we make, knowing what the projects are and who the teams are, is that those teams used an integrated design process to get an integrated design outcome. In other words, it's perfectly possible that they used a totally linear [traditional] approach and came up with a good design, but it's more likely that they had an integrated approach in order to get there.

The design teams that we work with are really good, they know how to put enough information into the drawings at an early stage so as they get later in design and especially in construction there's less effort, fewer change orders, and fewer [increases in] construction costs. So that would tell you that it's worthwhile to spend more money on design to save money in construction. Some of it comes down to what kind of project you have and what kind of fee a team is able to command [to be able to afford the extra initial costs of an integrated design process].

Cost Drivers for Green Buildings

What drives the costs of green buildings? First, let's look at factors relating to design costs. Key drivers include the following:

- Experience with LEED/high-performance projects
- Level of LEED certification required
- Team structure
- Design process and scope
- LEED documentation
- Design fees

TEAM EXPERIENCE

Team experience with LEED and high-performance projects is obviously a critical factor, for two reasons. First, teams with little or no experience will naturally assign a "risk premium" to their fees, hoping to get paid for their "learning curve" by the first client. Second, teams with experience have developed shortcuts, written standard specifications, researched alternative materials, and generally have learned to pick team members with similar experience. Therefore, they don't need to assign a risk premium to their fees and are more likely to treat LEED project requirements as "business as usual" rather than an added burden to their conventional design process.

Matthiessen comments on this issue, saying that teams need to get adequate fees to be able to do the early-stage studies that lead to more cost-effective outcomes.*

---

*Interview with Lisa Matthiessen, March 2008.

I would say that we're seeing an increasing number of projects now that have less sophisticated teams; with project types like really fast design-builds (development-type projects) these teams are having trouble with this notion because for them it's changing the way they work. With those projects where the clients are tough about fees during design, it's more difficult for the design teams to make a [process] change and actually spend more time during design in order to get a better result [during construction]. I've seen more commitment [to integrated design], but it's commitment that struggles with the notion [on the owner's part] that they have to pay more to get this.

LEED CERTIFICATION LEVEL

Level of LEED certification sought is clearly an issue. As you move to higher levels of LEED certification, even with an integrated design process, you are likely to add higher cost elements such as green roofs, photovoltaics, and certified wood products. You are also likely to want a larger number of studies in the design phase, including natural ventilation analyses, computational fluid dynamic studies, more frequent energy modeling, and the like. In some cases, nonetheless, we have examples of LEED Platinum being accomplished for zero or low cost premium, considering both design and construction costs, because teams find ways to cut capital costs by "right sizing" equipment, for example. (Read Leith Sharp's Foreword, for one such example.) For argument's sake, Table 7.1 presents estimates I have found for LEED project costs, including both design and construction. You can find studies with both higher and lower estimates, so please use these numbers only as a rough guide. The cost increases do not factor in the increased benefits, a subject tackled in Chap. 8.

TEAM STRUCTURE

One would expect that the more consultants there are on a team, the higher the design costs, since coordination will be an issue, as will the need to pay for expertise. Large numbers of consultants are typical on more complex projects such as performing arts centers, laboratories, and similarly complex buildings. Where an architect has an engineering function in house or where there are fewer consultants, there's an opportunity to save on design costs. However, a key element in team structure is the role of the

TABLE 7.1 RANGES OF LEED PROJECT COST PREMIUMS 2008 (AUTHOR'S ESTIMATES)	
LEVEL OF LEED CERTIFICATION	**OVERALL COST PREMIUM**
Basic certified	0% to 2%
Silver	1% to 4%
Gold	2% to 5%
Platinum	2% to 10%

general contractor and key subcontractors such as the mechanical contractor. When they understand how high-performance projects come together and are well-integrated into the building team from the beginning, cost premiums tend to evaporate. Indeed, without accurate estimates for the cost of mechanical systems, many energy-saving opportunities can be missed.

Matthiessen sees a link here between early contractor involvement and project costs.

Contractors are becoming much savvier. As they understand what this takes, they're not charging more for it. We're seeing more contractor-related points [in LEED-certified projects]. A lot of contractors have a much better understanding of those LEED points, are lot more willing to do them and are not asking for more money. Again, that's not necessarily changing the design, it's changing the construction practices and it is bringing costs down.

DESIGN PROCESS AND SCOPE

You can expect that a serious commitment to integrated design will add to design costs, because of additional meetings, design charrettes, and further studies and analyses during the design period. A typical all-day meeting is going to cost $20,000, if it involves many consultants, each billing $1200 to $2000 for the day. The least costly approach is to get everyone together just once, but not to leave the project venue until most key design decisions or directions are made. This approach is especially important when consultants are brought in from out of town or even out of country. You'll see that multi-day charrettes are often quite useful; you'll also see that shorter meetings can also work, in cases where the client side is really well informed about LEED and experienced in delivering high-performance projects. Other authors note that integrated design can require a series of facilitated charrettes.

It's also important to do most of the key thinking at the initial meeting and to use energy models particularly to get better and earlier design decisions. Some of the new developments in Building Information Modeling (BIM) promise to allow energy outcomes to be modeled for alternative design approaches very early in schematic design. At the USGBC's 2007 Greenbuild show, one BIM vendor showed a promising approach that would eventually allow modeling of approximate energy outcomes even with rough sketches.* Relating to the issue of getting to higher levels of sustainability without spending more, Matthiessen says,

Oddly enough, some of that is just using an energy model correctly, which we're seeing more teams doing. They're using an energy model earlier in the process as a design tool. They're using it as a way to make smart choices. Some of the changes they're making are not actual changes to the design, rather, the changes are to the operation of the building. They're looking at operating sequences and things like static pressures. They are looking at the assumptions they've made about how the building will be designed and operated and being a little tougher on those using the energy model [to provide specific advice and feedback on the energy implications of alternative designs].

---

*Plenary presentation by Phil Bernstein, Autodesk, at the USGBC's *Greenbuild* conference, Chicago, November 2007.

PLATINUM PROJECT PROFILE

Highland Beach Town Hall, Highland Beach, Maryland

The 2200-square-foot community center is located in Highland Beach, Maryland. Completed in the spring of 2006, the $500,000 Highland Beach Town Hall serves as a meeting and gathering space for community residents. Two-thirds of the roof is vegetated and can absorb up to 99 percent of a 1-inch rainfall. A ground source heat pump system is used for heating and cooling. Grid-tied photovoltaic panels are expected to produce 100 percent of the energy demand. Offsets purchased from American Wind Energy prevent 14,953 pounds of CO_2 from entering the atmosphere.*

LEED DOCUMENTATION COSTS

The requirement for LEED documentation could cost between $25,000 and $50,000 for team coordination and LEED project management services. Whether it's performed in-house at an architecture firm or done with outside consultants, there is a higher level of effort required to coordinate all the design team members and to keep the LEED aspects of the project on track. As LEED becomes fully integrated into design practice over the next half-decade, you can expect the costs for LEED project coordination, documentation and certification services to diminish, but not disappear altogether.

ADDED DESIGN FEES

Of course, building owners and developers don't want to pay higher design fees than they need to, but a standard approach is to select an architect for a high-performance project, then negotiate fees. This puts the owner at a disadvantage in fee negotiations, unless there are strong team-building activities and other methods early in design to reduce perceived risk by the architect. When selecting a "starchitect" (star architect), the owner is committing to higher design fees to get a high-design project.

My basic conclusion is that if building owners want high-performance green building design, they should be willing to pay what it costs to engage the very best practitioners. By the same token, as the owner, you should push very hard for the designers to figure out how to lower construction costs with their design choices. It's obviously worth paying higher design fees if you can get a less expensive overall result. Construction costs typically make up 92 percent of the total cost structure, so even adding 10 percent to design fees can be justified, if the designers can reduce overall construction costs by even 1 percent.

One factor leading to higher design fees is the role of specialized consultants on really complex projects. Every laboratory needs a lab consultant; every performing arts venue needs a theater consultant and a lighting consultant, and so on. Even large office buildings

---

*Highland Beach "Green" Town Hall [online], www.dnr.state.md.us/ed/pdfs/highlandbeach.pdf, accessed April 2008.

increasingly have specialized facade consultants and "climate engineers," a specialty pioneered by firms such as Arup and the German firm, Transsolar.* These specialties often come out of Europe because design fees for mechanical engineers are about one-third higher there than in the United States, about 2 percent of total construction as against 1.5 percent.[†] Therefore, more effort can be devoted to systems engineering rather than just to equipment specification. On this subject, Matthiessen says,

> Some teams want to bring in a lot of new experts that they didn't have before. Other teams are trying to do it internally and there are pluses and minuses to both approaches. One is not particularly better than the other. I think for most people integrated design means that they're including more of the traditional design team members earlier in the process. I have not seen that many projects where it's a whole new paradigm. It's more like they're improving the approach they already have.

PLATINUM PROJECT PROFILE

Home on the Range, Billings, Montana

Now a 10,000-square-feet office for the Northern Plains Resource Council and the Western Organization of Resource Councils, this building formerly housed a derelict grocery store. What was once a Billings eyesore is now a local landmark and has brought new life into a downtown portion. The total construction cost for the building was $1.4 million. A 10-kW grid-tied photovoltaic system generates approximately 37 percent of the building's electricity, and a solar thermal water system supplies all of the hot water. The building uses evaporative cooling and radiant heating for space conditioning. Composting toilets and a water-free urinal help the building reduce water use by two-thirds (compared to a similar building). Cubicles, wood trim, solid oak doors, biofiber boards, carpet tiles, bathroom wall tiles, fly-ash concrete, furniture upholstery, bathroom sinks, and kitchen tiles are examples of products containing recycled, salvaged, and sustainable materials.[‡]

Additional Cost Considerations

There are other potentially significant factors that determine what a LEED project will cost, on a "dollars per square foot" basis. When estimating a LEED project, these factors often determine the final project budget. Some of them are quite significant but may have nothing to do with the level of LEED certification or energy performance sought.

---

*See the work of Matthias Schüler of Transsolar at www.transsolar.com, accessed April 2008.
[†]Personal communication, Patrick Bellew, CEO of Atelier Ten, London, February 2008.
[‡]Home on the Range: An In-Depth Look at Montana's Greenest Commercial Building [online], http://www.greenhomeontherange.org, accessed April 2008.

THE INITIAL BUDGET

Low budget projects, such as design/build retail store or tilt-up construction suburban offices are going to be harder to LEED-certify without adding costs, because so many of the costs have already been optimized for that project type, particularly to reduce capital cost by not exceeding any code requirements. In looking at small retail stores, for example, I have found a 5 percent cost increase for the first certification. However, the USGBC's "volume build" program can be used to keep costs low, because it allows a retailer to certify a prototype, then just submit any site-related changes for review, for each subsequent store. By contrast, a project with a large budget per square foot can more easily absorb any added costs for LEED certification.

TIMING OF THE PROJECT

A significant determinant of cost is when the project is bid; in 2006 and 2007, for example, I've been told that some projects in major cities had trouble even getting sub-contractor bids, because everyone was just flat-out too busy. As the commercial building sector slows down in 2008 and 2009, as seems likely, contractors will likely be more eager for work, and prices should come down a little. Continuing inflation of construction materials is also a factor in adding cost to project budgets.

LOCATION OF THE PROJECT

In many cities, there are a lot of LEED projects. In those locations, contractors are becoming accustomed to installing green roofs, underfloor air systems, hydronic space conditioning, and similar departures from "normal practice," and don't have to include a "fear factor" in their bids. In other places, the first few LEED projects are probably going to have to work harder at subcontractor education and sourcing alternative materials, to keep costs down.

CLIMATE

Since many of the high-performance LEED projects have aggressive energy-savings goals, climate can be an issue in determining costs. For example, many projects on the West Coast, where humidity is not a significant design factor and temperatures are mild much of the year, can rely on 100 percent dedicated outside air systems (using economizer cycles) for much of their cooling during the year. Such systems are not really possible in the hot humid Southeast, and so cooling options may not be as plentiful. A study of laboratory buildings by Davis Langdon (using the Labs 21 EPC rating system) showed that the incremental cost of a Gold building could nearly double between coastal Santa Barbara, California and California's interior Central Valley.

DESIGN STANDARDS

Certain public and private institutions require higher levels of design than just a "code" building. Certain campus settings already provide parking; central steam power, chilled water, and/or electricity; open space; multiple transportation options; and other services that result in anywhere from 5 to 15 "embedded" LEED points toward certification for any project. Similarly, a shopping center developer with higher standards for site development could "embed" 6 to 12 LEED points for every retail store that wanted to become certified, without the store having to spend anything more.

PROJECT SIZE

You would expect smaller projects to have a higher cost premium for LEED certification, because certain of the costs of LEED (e.g., documentation assistance, energy modeling, and building commissioning) have fixed-cost elements independent of project size that will add to the cost per square foot. Above a certain size, perhaps 50,000 square feet for new projects or major renovations, this "size cost premium" starts to disappear. Let's say such a project has a budget of $5 million and let's say that the LEED "fixed cost" items for a Silver certification total $75,000. That's a cost premium of 1.5 percent, generally quite tolerable for most projects, assuming there are not too many other *net* cost additions from higher energy efficiency, for example. At $10 million, the fixed cost premium is likely below 1 percent.

FEASIBILITY OF LEED MEASURES

Another cost factor is the feasibility of LEED actions. For example, in most urban areas, recycling 75 percent of construction waste is virtually a no-cost item for the project. However, for projects in rural areas, there may be no construction waste recycling opportunities. To get the "lost" two points from some other LEED measures, for example, might add cost to an identical project, everything else held equal, if it wanted to qualify for the same certification level. The same would hold true for many of the materials and resources credits that might rely on a local supply chain for supply of recycled-content materials or certified wood products. (Ironically, more than a few of the LEED Platinum projects we profile in this book are located in rural areas.)

DESIGN PROCESS AND CREDIT SYNERGIES

Certain sustainability measures exhibit synergies that generate multiple LEED credits from one action. Take green roofs, for example. Green roofs not only reduce the urban heat island effect, but also mitigate stormwater runoff, create habitat and urban open space, conserve water use in landscaping, reduce building energy use, and may even qualify for an innovation credit in the LEED system. The same measure could provide for up to 8 LEED points, serving to offset the higher costs of green roofs, typically $10 to $20 per square foot, by eliminating other costs that would have been incurred to achieve equivalent benefits. Of course, in certain climates such as the hotter and drier western states, reflective roofs may represent a better approach.

PLATINUM PROJECT PROFILE

Vento Residences, Calgary, Alberta

Designed by Busby Perkins+Will, this 38,750-square-feet (3600-square-meter), three-story mixed-use project includes retail, underground parking, and 22 residences (including two affordable housing suites). A 45 percent reduction in energy derives from an enhanced building envelope; improved insulation levels; double-glazed, argon-filled, low-e windows; exhaust air heat recovery ventilation; and lighting occupancy sensors. The energy cost performance of the Vento Residences is 47 percent better than Canada's Model National Energy Code for Buildings. Dual-flush toilets, low-flow fixtures, and rainwater reuse contribute to reducing potable water use by 50 percent.*

Courtesy of Windmill Development Group.

Controlling Costs in LEED Projects

Architect Peter Busby, leader of the sustainable design practice at Perkins+Will internationally, has designed a number of LEED-certified projects. His approach to controlling costs involves several key elements:[†]

*Canada Green Building Council [online], http://my.cagbc.org/green_building_projects/leed_certified_buildings.php?id=83&press=1&draw_column=3:3:2, accessed April 2008.
[†]Personal communication, Peter Busby, April 2008.

- Have a clear green design goal from the outset.
- Make sure the design team is completely integrated.
- Incorporate green elements in the design from the beginning.
- Have centralized management of the green building process.
- Team members should have experience with/knowledge of green building.
- Obtain sufficient technical information to make informed decisions.
- Provide sufficient upfront time and funding for studies to get the technical information.
- Always insist on life-cycle costing of green investments.*

We will return to these points in several places in this chapter, since each design team has to address the challenge of identifying green building costs (and benefits) and justifying them to clients. (Chap. 6 presented the business case for green buildings by placing the full range of benefits into perspective, often a necessary prelude to considering whether to bear additional costs.)

High Performance on a Budget

A large developer-driven, build-to-suit project in Portland, Oregon, the Oregon Health & Science University's Center for Health and Healing, occupied in the fourth quarter of 2006, exposed flaws in the notion that dramatically higher levels of performance must always lead to significantly higher capital costs. The 412,000-square-feet, 16-story, $145 million project received a LEED Platinum rating early in 2007, the largest project in the world thus far to achieve this rating. The developer has reported a total cost premium, net of local, state, and federal incentives, of one percent.[†] The total costs for the mechanical and electrical systems were about $3.5 million below the initial budget estimates from the general contractor, based on standard approaches to space conditioning and lighting.[‡] These results were achieved with a strong bias toward integrated design, which the engineers define as having the same system perform multiple tasks. For example, an underground garage ventilation fan doubles as a smoke evacuation fan in case of a fire, through a motorized damper. At the same time, energy and water modeling indicated a 61 percent savings on future energy use and a 56 percent savings in water consumption. In other words, from a performance standpoint, this project demonstrates the benefits of an integrated design process, coupled with an experienced developer and design team willing to push the envelope of building design.

One of the key strategies employed in integrated design projects is the notion of "cost transfer," particularly from mechanical and electrical systems, to the rest of the project,

---

*A fuller discussion of life-cycle costing is far beyond the scope of this book. For one example, see www.wbdg.org/resources/lcca.php, accessed April 2008.
[†]Personal communication, Dennis Wilde, Gerding Edlen Development, 2006.
[‡]"Engineering a Sustainable World" by Interface Engineering, October 2005, available at www.interfaceengineering.com, accessed April 2008.

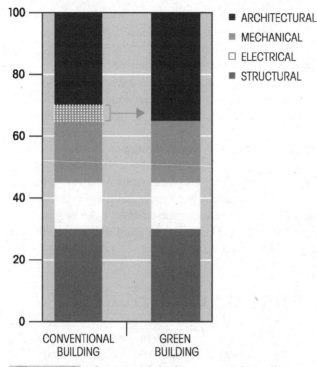

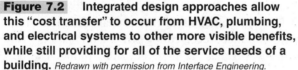

Figure 7.2   Integrated design approaches allow this "cost transfer" to occur from HVAC, plumbing, and electrical systems to other more visible benefits, while still providing for all of the service needs of a building. *Redrawn with permission from Interface Engineering.*

as illustrated in Fig. 7.2. Here's a good example, the OHSU project employs shading devices over the south-facing windows on the fourth through the fourteenth floors of the 16-story high rise. The east and west facades contain few windows, only stairways, and are naturally ventilated using a "stack effect" approach. As a result, there is much less solar heat gain in the summer months, with an attendant cost reduction in the required tonnage of air-conditioning. The cost savings from the reduced air-conditioning system size more than pay for the south-facing overhangs and produce a cost transfer to other aspects of the building, such as a large green roof and 60-kW of building-integrated photovoltaics, with panels integrated onto the south-facing sun shades.

The more developers engage experienced green design and construction firms, the more they require their consultants to produce high-performance results (without excuses), the more likely it is that overall project costs will not exceed costs for a conventional project that does not provide the benefits of a high-level green building.

Many of the green building measures that give a building its greatest long-term value—for example, onsite energy production, onsite stormwater management and water recycling, green roofs, daylighting, and natural ventilation—often require a higher capital cost. However, many project teams are finding that these costs can be paid for by avoiding other costs, such as stormwater and sewer connection fees, or by using local utility incentives, state tax breaks, and federal tax credits. The key message from these projects is that integrated design also needs to look at "integrated costing," that is, considering the entire project budget or the construction budget.

While it is often possible to get a LEED-certified (and sometimes LEED Silver) building at no additional cost, as building teams try to make a building more sustainable, cost increments often accrue. This is especially true when the building owner or developer wants to showcase their green building with more expensive (but visible) measures such as green roofs or photovoltaics for onsite power production, or where there is a strong commitment to using green materials such as certified wood.

Summary of Cost Influences

Chapter 6 discussed the many business case benefits of green buildings, and made the points that costs are real, occur first, and must be justified to various stakeholders. Benefits are generally long-term, and costs are immediate, so many people tend to shy away from anything that will add costs, no matter what the potential benefits.

Table 7.2 shows some of the elements of green building design and construction decisions that may influence project cost. From this table of "cost influencers," you can see

TABLE 7.2 COST INFLUENCERS FOR GREEN BUILDING PROJECTS	
COST INFLUENCER	**POSSIBLE COST INCREASES**
1. Level of LEED certification sought	Zero for LEED-certified to 1–2 percent for LEED Silver, up to 4 percent for LEED Gold
2. Stage of the project when the decision is made to seek LEED certification	After 50 percent completion of design development, things get a lot more costly to change
3. Project type	With certain project types, such as science and technology labs, it can be costly to change established design approaches; designs for office buildings are easier to change
4. Experience of the design and construction teams in sustainable design and green buildings	Every organization has a "learning curve" for green buildings; costs decrease as teams learn more about the process
5. Specific "green" technologies added to a project, without full integration with other components	Photovoltaics and green roofs are going to add costs, no matter what; it's possible to design a LEED Gold building without them
6. Lack of clear priorities for green measures and lack of a strategy for including them	Each design team member considers strategies in isolation, in the absence of clear direction from the owner, resulting in higher costs overall and less systems integration
7. Geographic location and climate	Climate can make certain levels of LEED certification harder and costlier for project types such as labs and even office buildings.

that there is no right answer to the question: "how much does a green building cost?" I often tell audiences that the definitive answer to this question is simple—it depends!

Overall, costs associated with green design and construction may exceed 1 percent of construction costs for large buildings and 5 percent of costs for small buildings, depending on the measures employed.

Higher levels of sustainable building (for example, LEED Silver, Gold, or Platinum standard) may involve some additional capital costs, based on case studies of completed buildings in the U.S. LEED projects also incur additional soft (nonconstruction) costs for additional design, analysis, engineering, energy modeling, building commissioning, and certification. For some projects, additional professional services, for example—including energy modeling, building commissioning, additional design services, and the documentation process—can add 0.5 to 1.5 percent to a project's cost, depending on its size.

Green Building Cost Studies

Given the high level of interest in the costs of green, it's surprising that there are so few rigorous studies of the cost of "green versus nongreen," for similar projects. Here are a few of the studies that point the way toward a better understanding of green building costs.

THE 2003 CALIFORNIA STUDY

A 2003 study for the State of California was the first rigorous assessment of the costs and benefits of green buildings.* Drawing on cost data from 33 green building projects nationwide, the report concluded that LEED certifications add an average of 1.84 percent to the construction cost of a project. For Gold-certified office projects, construction cost premiums ranged from 2 to 5 percent over the cost of a conventional building at the same site.

THE 2004 GSA COST STUDY

A 2004 study for the General Services Administration of the costs of achieving various levels of LEED certification for government buildings looked at both new construction and remodeling projects. It supports somewhat similar conclusions to the work for the State of California. For example, in the California analysis, a $40 million public building seeking a LEED Gold level might expect to budget about 2 percent, or $800,000, extra to achieve this certification.

The 2004 study carefully detailed two typical projects, a new federal courthouse (with 262,000 square feet and a construction cost of $220 per gross square foot) and

---

*Gregory Kats et al., *The Costs and Financial Benefits of Green Buildings*, 2003, available at www.cap-e.com/ewebeditpro/items/O59F3303.ppt#1, accessed March 6, 2007.

an office building modification (with 307,000 square feet and a construction cost of $130 per gross square foot). At that time, the study estimated the additional capital costs of both types of GSA projects ranged from negligible for LEED-certified projects up to 4 percent for Silver level and 8 percent for Gold level.*

PLATINUM PROJECT PROFILE

Science & Technology Facility at the U.S. Department of Energy's National Renewable Energy Laboratory (NREL), Golden, Colorado

The U.S. Department of Energy's National Renewable Energy Laboratory (NREL) is a multi-story facility that houses NREL's solar and hydrogen research groups. Designed by SmithGroup and completed in June 2006, the 71,000-square-feet facility cost $22.7 million. Designed to reduce overall energy use by 40 percent, the project employs daylighting for 100 percent of the ambient light requirements in the offices. An energy recovery system is used for ventilation in the labs. Eleven percent of the materials used are from recycled products, and 27 percent of the construction materials were manufactured within 500 miles of the building site.[†]

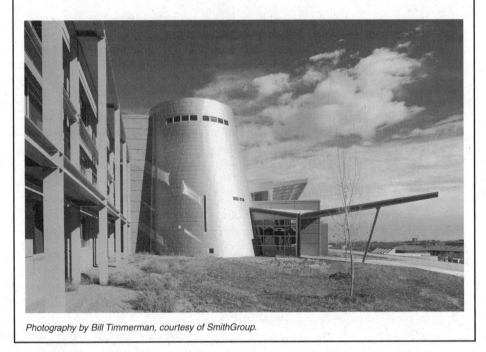

Photography by Bill Timmerman, courtesy of SmithGroup.

*Steven Winter Associates, "GSA LEED Cost Study," downloadable (578 pages) from the *Whole Building Design Guide* web site, www.wbdg.org/ccb/GSAMAN/gsaleed.pdf, accessed March 18, 2007. The authors note: "The construction cost estimates reflect a number of GSA-specific design features and project assumptions; as such, the numbers must be used with caution [and] may not be directly transferable to other project types or building owners" (*ibid.*, at p. 2).

[†]NREL, http://www.nrel.gov/news/press/2007/507.html; www.nrel.gov/features/07-04_science_tech_facility. html; www.eurekalert.org/features/doe/2007-04/drel-st040507.php, accessed April 22, 2008.

GREENING AMERICA'S SCHOOLS

In 2006, Gregory Kats released a study of the costs and benefits of greening K–12 schools. The report, "Greening America's Schools" became one of the most important documents to justify green buildings for a very large market segment, K–12 schools.* In Chap. 6, we profiled the benefits of green schools outlined in that report. The report studied 30 green school projects in 10 states, completed from 2001 through 2006, and concluded that the average green cost premium was 1.7 percent, or about $3 per square foot. As defined by the report, the "green premium" is the "initial extra cost to build a green building compared to a conventional building." Typically this cost premium is a result of more expensive (and sustainably sourced) materials, more-efficient mechanical systems, and better design, modeling, and integration, along with other high performance features. Many school architects use a state or school district's predetermined budget (or historical costs) as their metric for appropriate school cost. Some green schools have been built within the same budget as conventional schools, but many do need to spend extra money because of various design constraints. The data on costs as well as savings compared to a conventional design were generally supplied by the schools' architects.

PLATINUM PROJECT PROFILE

Sidwell Friends School

The Sidwell Friends School is a pre-K through 12th-grade Quaker independent school in Washington, D.C. Designed by Kieran Timberlake Associates, the three-story, 72,200-square-feet building was completed in September 2006. A constructed wetland treats and recycles wastewater for reuse in toilets and the cooling tower. Sidwell's green roof retains some stormwater on the roof and allows it to transpire back into the atmosphere. An onsite central energy plant serves the entire campus. Photovoltaic panels provide 5 percent of the building's electrical demand. Solar-ventilation chimneys, operable windows, and ceiling fans minimize the need for mechanical cooling.[†]

THE DAVIS LANGDON COST STUDIES

We gave an example earlier in this chapter of the Davis Langdon project cost study. In 2004, the firm's first LEED cost study offered strong evidence, based on 94 different building projects of vastly different types, that the most important determinant of project cost is not the level of LEED certification sought, but rather other more conventional issues such as the building program goals, type of construction, and the local construction economy. In this study, the authors concluded that there was

---

*Capital-E [online], "Greening America's Schools, Costs and Benefits," October 2006, www.cap-e.com/ewebeditpro/items/O59F11233.pdf, accessed April 26, 2007.

[†]www.sidwell.edu/about_sfs/greenbuilding.asp; www.kierantimberlake.com/pdf_news/sidwell-friends-school.pdf; www.aiatopten.org/hpb/overview.cfm?ProjectID=775; accessed April 22, 2008.

no statistically significant evidence that green buildings cost more per square foot than conventional projects, primarily because so many factors influence the cost of any particular type of building.* Based on these results, one would expect more pressure from owners and developers for design and construction teams to aim for high LEED goals, because these buildings are indeed perceived to offer higher value for the money spent.

The study's authors commented, "From this analysis we conclude that many projects achieve sustainable design within their initial budget, or with very small supplemental funding. This suggests that owners are finding ways to incorporate project goals and values, regardless of budget, by making choices. However, there is no one-size-fits-all answer. Each building project is unique and should be considered as such when addressing the cost and feasibility of LEED. Benchmarking with other comparable projects can be valuable and informative, but not predictive."

The 2006 follow-up report by Davis Langdon on 130 projects reported these conclusions: Most projects by good design teams have "embedded" 12 LEED points (out of 26 needed for certification) and most could add up to 18 points to achieve basic LEED certification with minimal total cost, through an integrated design approach.[†] Of 60 LEED-seeking projects analyzed, more than half received no supplemental budget to support sustainable goals. Of those that received additional funding, the supplement was typically less than 5 percent, and supplemental funding was usually for specific enhancements, most commonly photovoltaics. In other words, the results of this study indicate that any design team should be able to build a LEED-certified building at no additional cost, and a LEED-Silver building with only a minor cost increase.

COSTS OF GREENING RESEARCH LABS

Davis Langdon also studied the impact of climate on the costs of a research lab. Costs ranged from 2.7 to 6.3 percent premium for a LEED Gold project, and 1.0 to 3.7 percent for a LEED Silver project (the study assumes the same design was constructed in various cities at the same time).

The key cost message to owners and developers (and design and construction teams) is that sustainability needs to be a "program" issue, that is, it needs to be embedded in the goals of the project and not treated as an add-on cost element. *This conclusion is not just a matter of semantics; it goes to the very heart of the question, "what is the purpose of this building or project?" If sustainability is not a core purpose, then it's going to cost more; if it is essential to the undertaking, then costs will be in line with non-green buildings of the same type.*

---

*Lisa Matthiessen and Peter Morris, "Costing Green: A Comprehensive Database," Davis Langdon, 2004, available at www.davislangdon.com/USA/research. The 2006 update, "The Cost of Green Revisited," can be found at www.davislangdon.com/USA/Research/ResearchFinder/2007-The-Cost-of-Green-Revisited, accessed April 22, 2008.

[†]U.S. Green Building Council, November 2006, LEED Cost Workshop.

> ## PLATINUM PROJECT PROFILE
>
> ### Hawaii Gateway Energy Center, Kailua-Kona, Hawaii
>
> Designed to house research, development, and demonstration facilities for the Natural Energy Laboratory of Hawaii, the Hawaii Gateway Energy Center was completed in January 2005. The 3600-square-feet facility's total project cost was $3.4 million.* Designed to be a thermal chimney, the building captures heat and creates air movement using only the building's form along with thermodynamic principles. The project has a ventilation rate of 10 to 15 air changes per hour without the use of a mechanical system. A 20-kilowatt photovoltaic system provides all the energy needs. The building was designed to be 80 percent more energy efficient than a comparable facility built to ASHRAE 90.1-1999 standards.[†]

SOFT COSTS FOR GREEN BUILDING PROJECTS

The 2004 GSA study mentioned earlier also looked at "soft costs," costs for things that are not part of building construction. The study estimated soft costs for additional design and documentation services ranged from about $0.40 to $0.80 per square foot (0.2% to 0.4%) for the courthouse and $0.35 to $0.70 per square foot (0.3% to 0.6%) for the office building modernization project. One caution: the added percentage of total cost may be higher for smaller projects.

Therefore, each building team should look at every cost that a project would incur, from permitting and site development to furniture and fixtures, before deciding that a particular green measure is "too costly." Integrated design requires systems thinking, to avoid the near universal tendency to look at individual cost items in isolation, a process euphemistically called "value engineering." Deciding which costs are going to provide the highest value in a given situation should be a primary task of the architect, working in concert with the client, the building owner or developer, and the builder.

One thing is certain: there are costs associated with green building projects that need to be taken into account, especially with those aiming at LEED certification. Many projects do not consider these costs especially onerous, but some do. Table 7.3 shows some of the potential "soft" costs, that is, those not construction related. Some of these costs should be considered essential to good project design and execution, specifically building commissioning and energy modeling, while others are more clearly associated with the LEED certification effort.

---

*This is obviously a high-cost project owing to the many special features included.

[†]U.S. Green Building Council, http://leedcasestudies.usgbc.org/overview.cfm?ProjectID=592, accessed April 2008.

TABLE 7.3 "SOFT COSTS" OF LEED CERTIFICATION, 2008*		
ELEMENT	**COST RANGE**	**REQUIRED IN LEED?**
1. Fundamental building commissioning	$0.40 to $1.00 per sq ft, $20,000 minimum	Yes
2. Energy modeling	$15,000 to $30,000	Yes; depends on size and complexity
3. LEED documentation	$25,000 to $90,000	Yes; depends on complexity of project, team experience and level of certification
4. Eco-charrettes	$10,000 to $20,000	No
5. Natural ventilation modeling	$7,500 to $20,000	No
6. Enhanced commissioning services	$3,000 to $15,000	No
7. Daylighting design modeling	$3,000 to $10,000	No (some utilities offer this as a free service)
8. Measurement and verification plan (for LEED credit EA5)	$10,000 to $30,000	No, an optional credit point

*Based on the author's professional experience.

Integrated Design Can Reduce Costs

Often, the traditional "design-bid-build" process of project delivery works against the development of green buildings. In this process, there is often a sequential "handoff" from the architect to the building engineers to the contractor, so that there is a limited "feedback loop" arising from the engineering aspects of building operating costs and comfort considerations back to basic building design features.

In a standard design process, for example, the mechanical engineer is often left out of the architect's building envelope design considerations, yet those decisions are often critical in determining the size (and cost) of the HVAC plant, which can often consume up to 15 percent or more of a building's cost. Along the way, the standard "value engineering" exercise often involves reducing the value of the HVAC systems by specifying lower efficiency (cheaper) equipment, possibly reducing the R-value of glazing and insulation, and other measures that reduce first costs, but higher operating costs for energy supply for the lifetime of the building. (Lifetime operating costs are typically 80 percent or more of a building's total costs.)

PLATINUM PROJECT PROFILE

Shangri La Botanical Gardens & Nature Center, Orange, Texas

Near the border of Texas and Louisiana sits the 250-acre Shangri La Botanical Gardens and Nature Preserve. Closed after a snowstorm wiped out most of the gardens, the new Shangri La Botanical Gardens center was designed by Lake/Flato architects and reopened in March of 2008. The building houses research facilities, an outdoor education center, classroom pavilions, and a visitors' center. Solar panels throughout the property produce 21 percent of the center's energy needs. Many of the site's structures were constructed out of reclaimed brick from a 1920-era warehouse; the asphalt parking lot was salvaged during the repaving of one of Orange's city streets. Cypress trees recovered from a Louisiana river bottom were milled for use as siding, slat walls, fencing, doors, and gates. Rainwater is collected and used in toilets and for irrigation.*

Flack & Kurtz' Dan Nall comments on how this takes place.[†]

The stuff that we specify—mechanical, electrical, plumbing, fire protection, etc.—typically represents anywhere from the high 20 percent up to 40 percent of the total construction cost of a building depending on the type of building. If it's a hospital or something like that, it might well be 40 percent. Certainly if it's a data center it could be that much or more.

We specify a lot of things that the owner is going to be buying, but we don't have quite as much leeway in managing our costs because a lot of things are either mandated by code, mandated by the functionality of the building, or somehow are not discretionary. Yes, we can economize on quality and look for bargain materials and components. But we also have to be very careful, because there are liability issues there also. We want to make sure that the thing lasts—that it doesn't break in a short period of time and we get sued. After all, we are specifying a significant fraction of the building's cost.

However, when the value engineering effort comes around, often times we're stuck with the issue of how much we are willing to compromise. We originally put this stuff in the design for a reason. Very likely it wasn't because we liked the way it looked. Some could make the charge that in some cases (and in some cases I'm sure it's quite correct) the reason we drew what we drew, designed what we designed, and the reason we specified what we specified was an over-abundance of caution. Surely, that does happen, and as engineers we need to recognize when we're being overly

---

*Sources: www.architechmag.com/news/detail2.aspx?contentID=58301495; and www.aia.org/aiarchitect/thisweek08/0321/0321d_shangrila.cfm, accessed April 22, 2008.
[†]Interview with Dan Nall, March 2008.

cautious and to be able to pull an element out of the project if necessary, to help adjust the cost. But it is, from time to time, an agonizing kind of decision to make.

Similarly, I'm sure on the architectural side, when asked to value engineer, the architects is continually agonizing over the extent to which making a particular substitution or doing something differently compromises the architectural vision to the point that it's no longer valid. We're both agonizing in our own ways about how much we have to compromise what we think the project needs in order to meet the budget. Of course, it's really tough because every project that I'm working on right now is over budget, because construction prices are rising at a fairly rapid rate. I believe that a lot of building owners are not really realistic about the extent to which this escalation is going on and they have, as people say, eyes that are bigger than their wallets.

With the necessity to cut initial costs to remain within budgets, the conclusion is that key design decisions are often made without considering long-term operating costs. Most developers, building owners, and designers find that a better process for creating green buildings involves an integrated design effort in which all key players work together from the beginning. Developers and owners have discovered cost savings in building design and construction through the use of integrated design approaches as well as other "nontraditional" measures, which might include bringing in the general contractor and key subcontractors earlier in the process to help with pricing alternative approaches to achieve required comfort levels in a building. Integrated design has one other major benefit. It makes it nearly impossible to "value engineer" too much out of the building. For example, adding south-facing external shading devices adds cost. When asked to remove these from the design, the engineer might respond with a request for more HVAC capacity, and that might cost even more than the reduction in shade cost. Hence, the shading will stay in the design.

PLATINUM PROJECT PROFILE

Child Development Centre, University of Calgary, Calgary, Alberta

Completed in August 2007, the Child Development Centre houses a child care center, the Calgary Health Region offices, researchers, clinicians, and community practitioners. The 125,000-square-feet, four-story building cost $23 million to build. Designed by Kasian Architecture, the facility was designed to reduce energy costs by over 70 percent and reduce water use by 55 percent. A wall-mounted photovoltaic system is capable of producing 65,000 kilowatt-hours of electricity per year. Additional sustainable, high-performance features include heat recovery wheels, FSC-certified wood, renewable cork flooring, dual-flush toilets, water-free urinals, radiant cooling panels, and operable windows.*

---

*Bradley Fehr, University of Calgary Child Development Centre Opens: LEED-Platinum-certified building sets Canadian and world records, Journal of Commerce, November 12, 2007 [online], http://www.joconl.com/article/id25047, accessed April 2008. Kasian Architecture Interior Design and Planning Ltd (October 22, 2007). "Kasian designs LEED Platinum certified building." Press release. Retrieved April 2008.

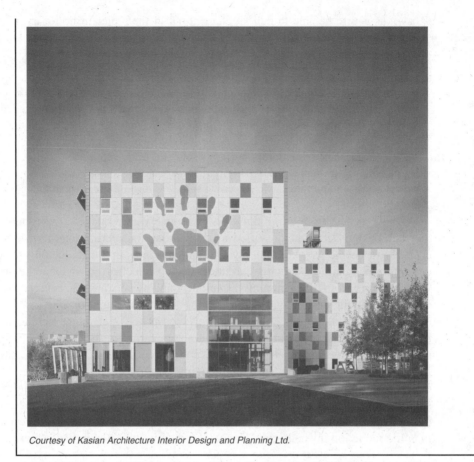

Courtesy of Kasian Architecture Interior Design and Planning Ltd.

Gross Costs and Net Costs

Because architects and engineers tend to think like design professionals and not owners, they often overlook the difference between gross costs and net costs in determining whether to accept or reject sustainable design measures, to the detriment of the project. For example, underfloor air distribution systems cost more, typically $4 to $7 per square foot, on a gross basis. However, by bringing power and data cabling directly to each work station, they eliminate the need for systems furniture that is pre-wired. This furniture can cost, let's say $1000 (or more, compared with standard partitions) per 150 square foot (gross area) work station, a cost of $6.67 per square foot. The problem is that furniture is in the "FFE" budget and not the base building budget, so this potential cost offset often gets overlooked in early design considerations, even though in many cases the same building owner is paying for both costs. A good example of "penny-wise and pound-foolish," don't you think? Here's another example: rainwater harvesting and reuse not only can contribute 6 to 8 LEED points (similar to the green roof example above), but also can reduce storm sewer system connection fees or "impact fees" by reducing or eliminating stormwater outflows from the site. Again, the design team or the owner may overlook these savings and rule out a rainwater collection and treatment system for cost reasons and, by so doing, actually add cost to the project! These are not far-fetched examples; I've seen them happen on several projects.

"PAYBACK" VERSUS "RETURN ON INVESTMENT"

Engineers (and to some extent, architects) are often unable to present economic information to higher-level decision-makers who prefer to speak the language of business, not design. A good example occurs in the world of energy savings. For example, suppose it cost $300,000 extra to secure annual energy savings of $100,000. This is a typical circumstance. An engineer might say that this measure has a "3-year payback" ($300,000/$100,000), in terms of how long it takes to recover the initial investment from the annual energy cost reduction. (This is what they remembered from the one course in "engineering economics" they had to take in college.) A more business-oriented person would say that this measure has a "33 percent, inflation-protected return on investment," because payback is measured in terms of today's energy costs, which are quite likely to increase in the future. Which approach sounds more inviting—waiting 3 years just to get your money back, or making a very high-return investment? Both use the same data, but one is more likely to get approval, don't you think?

Let's take the same situation into the world of commercial real estate. Commercial properties are typically valued as a multiple of "net operating income," typically determined by dividing the income by a capitalization rate expressed as a percentage (think of how a corporate bond is valued; it's the same approach). If I reduce annual energy costs by $100,000, a typical "cap rate" of 6 percent would yield an incremental increase in value of $1.67 million ($100,000/0.06). So the same investment in energy efficiency (in a commercial situation) would create a 566 percent immediate return on investment ($1,667,000/$300,000)! Quite a difference, don't you think, between a 33 percent return on investment (which of course sounds pretty good) and a 533 percent immediate increase in value! If engineers and architects learned how their clients make and talk about money, it would make it far easier to "sell" sustainable design investments for many projects. Again, the basic message is that design and construction professionals need to get out of their "comfort zone," to become more effective advocates for high-performance buildings.

8

INTEGRATED PROJECT MANAGEMENT—COST/BENEFIT ANALYSIS OF GREEN BUILDINGS*

Since 2004, the green building movement has hit its stride. Green building principles have driven massive innovation in product and materials development and building design, and a fundamental change in the process of how buildings are built, called *integrated design*, has helped builders to deliver on the green promise. Green buildings are energy efficient, water stingy, filled with daylight and healthy air. As such, they create environments where the building users are healthier and more satisfied, driving productivity gains, higher test scores in schools, lower absenteeism, and higher operational efficiency.

So why don't more people build green buildings? As we have discussed throughout this book, there are three main issues:

1 Many owners and designers don't want to change their current development, design, engineering, and/or construction processes and procedures.
2 Many don't understand the full range of costs and benefits inherent in making the changes required by a commitment to green development, design and construction practices, so they tend to shy away from trying new approaches.
3 Many don't understand the life-cycle cost benefits that result from implementing green building practices within the context of their project requirements.

With its widespread impact on the building industry, green building is now entering its *posthoneymoon phase*. It's becoming increasingly difficult for a qualitative review of green building benefits to influence owners and developers to blindly accept all green building initiatives. Most project teams face "cost versus benefit" questions for

---

*An original draft of this chapter was contributed by Paul Shahriari, Principal, GreenMind, Inc., www.greenmind. com. Paul is the creator of the *Ecologic3* LEED Project Management Software, www.ecologic3.com, accessed July 31, 2008.

the various measures proposed for their green building projects. Without a solid quantifiable set of metrics to assess a project, teams inevitably begin to see environmental initiatives take a back seat to traditional project forces such as schedule constraints and budget-related issues. I believe that with a well-established framework of vision, goals and life-cycle financial benefits, project teams will be able to better understand the owner's needs, and owners will be able to make better data-driven decisions.

In this chapter, I will focus on the fact that benefiting from the paradigm of green building requires all project stakeholders to understand the goals for which they are striving. In addition, there needs to be in place a project management process that addresses the full range of life-cycle financial benefits. As I've pointed out in this book, the foundation of green building success is the *integrated design process*. Simply put, the entire team needs to utilize the talents of all disciplines to accurately understand the impacts of green buildings. This chapter presents several examples of LEED project credit analysis and offers an introduction to a cost/benefit methodology that is critical to the success of any green project. At the end of the chapter, I examine a situation when the individual solutions are combined into a complete project analysis. In this case, the aggregate impact of both soft and hard costs can be analyzed immediately against the long-term benefits of various levels of LEED certification.

Introduction to the Environmental Value-Added Method

Traditional project delivery methods do not provide a very easy path toward the goal of successfully delivering a green building project. As you've learned from various examples presented earlier in this book, when evaluating the cost/benefit components of a green building, the starting point on all projects needs to be a visioning session among project stakeholders. This session should be a focused meeting that allows the project owner or developer to develop a set of goals, that the project design and construction team will refer to frequently during the design development process.

In the visioning session, the owner should bring the entire stakeholder team together to help participate in the visioning for the project. Sustainability and its application to green building can be defined by the concept of the Triple Bottom Line. The Triple Bottom Line concept is based on concerns in three main areas of impact: planet, profit, and people. Others define the impact areas as environmental, economic, social. If the owner can clearly define what elements of the triple bottom line they are striving for, then the team can more clearly articulate a design path to reach those goals.

Many times, project teams start a project with a LEED checklist, but no specific areas of focus. This can lead a team to "shop" for points, in some cases implementing elements that the owner might not have that much interest in, while leaving tougher credits aside. The owner should define for the project team (and for the project itself) the elements of the Triple Bottom Line that they are striving for. While defining the goals that the project should strive for, project teams should also define any constraints that might affect those areas of interest. Skipping this visioning session leaves most green building projects at the mercy of the dreaded "value engineering" or merely cost-cutting exercises, Table 8.1 presents a set of goals articulated during an actual project visioning session for a corporate campus project.

Once the team defines the goals strived for along with the constraints, the team should then evaluate the list for synergies among the Triple Bottom Line elements. Synergies can always be found. For example, in Table 8.1, under "Planet: Reduce Energy Consumption" and "Reduce Greenhouse Emissions" are linked to each other. These elements align with "Reduce Energy Costs" under the Profit category, which then aligns with the People category element of "Be a Good Corporate Citizen" and "Reduce Greenhouse Emissions." When multiple elements across multiple categories align, the team can be fairly sure that the ownership team will stay focused on the elements of high performance green design that deliver those elements (Table 8.10).

After this analysis, the team has an understanding of the benefits that the owner is striving for, and the team can begin designing the project. The benefits that the owner has chosen to focus on should become part of the EVA (*Environmental Financial Value Analysis*) log. Unlike VA (Value Analysis) or VE (Value Engineering) the EVA log is meant to retain sustainable elements in the building that have an important role in delivering the financial results the client wants in a given project.

TABLE 8.1   TRIPLE BOTTOM LINE GOALS FOR A PROJECT VISIONING SESSION FOR A CORPORATE CAMPUS

PLANET/ENVIRONMENTAL	PROFIT/ECONOMIC	PEOPLE/SOCIAL
GOALS TO STRIVE FOR		
Reduce energy consumption 50%	Reduce energy costs 50%	Be a good corporate citizen
Reduce greenhouse gas emissions 50%	Reduce water costs 50%	Provide a healthy work environment
Reduce water usage 50%	Reduce maintenance costs	Reduce greenhouse gas emissions
Reduce waste produced during construction and during operations	Increase productivity	Maximize utilization of resources
Protect biodiversity	Reduce risk of sick building-related issues	Reduce overall carbon footprint
CONSTRAINTS		
Site is already selected	Owners payback targets are <10 years	Limited experience internal to owners' team

LEED Rating System and EVA

The LEED rating system provides teams with a wonderful framework of green building focal points. Each credit's performance standards lead to Triple Bottom Line results. This chapter uses the LEED-NC rating system and examines several credits to analyze how this process applies to actual project decision making. The chapter in particular analyzes the following LEED credits:

- Sustainable Sites
  - Credit 6: Stormwater Design
  - Credit 7.2: Heat Island Effect; Roof
- Water Efficiency
  - Credit 1: Water Efficient Landscaping
  - Credit 3: Water Use Reduction
- Energy & Atmosphere
  - Credit 1: Optimize Energy Performance
  - Credit 2: Onsite Renewable Energy
- Indoor Environmental Quality
  - Credit 3.1: Construction IAQ Management
  - Credit 8.2: Daylight & Views; Daylighting

When evaluating credits for possible inclusion on a project, the team should consider the following impacts:

- COST IMPACTS
  - Soft Costs (Will additional design, engineering, or consulting be needed?)
  - Hard Cost (Will additional construction-related elements or services be needed?)
- TRIPLE BOTTOM LINE IMPACTS—EVA log
  - Planet/Environmental (Which elements does this credit support?)
  - Profit/Economics (Which elements does this credit support?)
  - People/Social (Which elements does this credit support?)

Let's start now with the analysis and see how this method helps generate alternatives and assist with decision making. We'll also look at the cumulative cash flows of each alternative, considering both costs and benefits.

This approach is not unique to this book. For example, Dan Heinfeld of LPA, says:*

We created a proprietary software program that we use for all of our projects and it has become an essential tool in the sustainable charrettes that we conduct. It follows the LEED format. The software creates a living document that assigns responsibilities, costs, points taken, points not taken, and further study elements using an online process. The document gets revised and updated at all stages of the project and in

---

*Interview with Dan Heinfeld, LPA, Inc., March 2008.

every charrette. It's also keyed back to the USGBC Reference Guide, so in that meeting we can just call it right up online. We have found that this tool is also the very beginning of the LEED documentation. We use that for every project here whether it's aiming for LEED or not.

The other thing that's different these days is that the people who are at the table are different. Number one, they are a lot more experienced [with green buildings] and number two, they are a lot more inclusive in terms of disciplines which allows for much better design integration. Also, we have found that the team can look a project more holistically that way because everyone is represented at the table. It absolutely helps reduce costs also because it's more integrated and there are more synergies. We believe sustainability is about added value, not added costs and we are proving that everyday here with our projects.

So, the essential goal of the LEED rating system and of integrated design is to achieve Triple Bottom Line benefits, while controlling costs through the Environmental Value Added process. Let's look now at how this approach plays out in analyzing individual design decisions, using the *Ecologic3* software to provide the analytical tool and comparison values.

SUSTAINABLE SITES: CREDIT 6—STORMWATER DESIGN

In this analysis, we see immediately that the solution of bioswales has dramatically higher payoffs than using pervious paving and that the payoff is positive from the beginning (Fig. 8.1, Table 8.2).

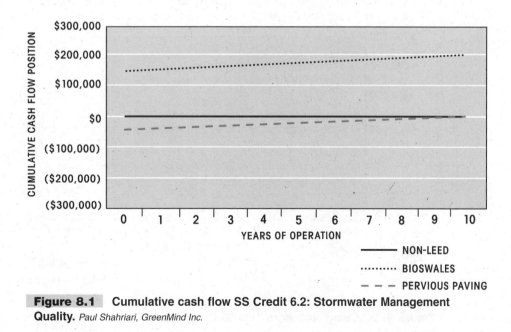

Figure 8.1   Cumulative cash flow SS Credit 6.2: Stormwater Management Quality. *Paul Shahriari, GreenMind Inc.*

TABLE 8.2   SUSTAINABLE SITES: CREDIT 6 STORMWATER DESIGN

	NON-LEED BUILDING	LEED CREDIT SOLUTION #1 BIOSWALES	LEED CREDIT SOLUTION #2 PERVIOUS PAVING
Soft Cost Impacts	None	Landscape architect and civil engineer collaboration	None
Hard Cost Impacts	None	Reduced land purchase for stormwater mitigation $150,000	1. Pervious paving utilized in lieu of standard asphalt paving + $200,000 2. Reduced land purchase for stormwater mitigation $150,000
Life-Cycle Benefits	None	Maintenance cost reduction $10,000/year	Maintenance cost reduction $10,000/year

SUSTAINABLE SITES: CREDIT 7.2—HEAT ISLAND EFFECT, ROOF

In mitigating the urban heat island effect, two obvious solutions are to design a LEED-compliant, high-emissive reflective roof or to install a green roof. Both alternatives have an up-front design and installation cost (Table 8.3), but the reflective roof achieves payback in the fifth year, while we have to wait until the ninth year to get a return on the green roof investment (Fig. 8.2). But, we may not want

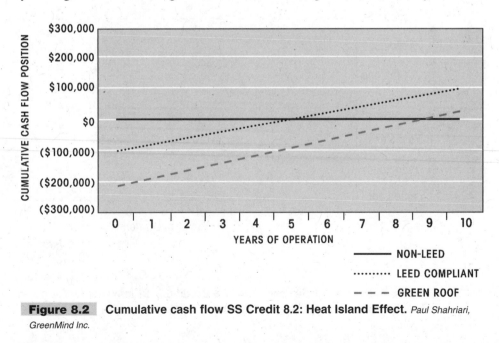

Figure 8.2   Cumulative cash flow SS Credit 8.2: Heat Island Effect. *Paul Shahriari, GreenMind Inc.*

TABLE 8.3 SUSTAINABLE SITES: CREDIT 7.2 HEAT ISLAND EFFECT, ROOF			
	NON-LEED BUILDING	LEED CREDIT SOLUTION #1 LEED COMPLIANT ROOF	LEED CREDIT SOLUTION #2 GREEN ROOF
Soft Cost Impacts	None	None	Structural engineering: additional design $10,000
Hard Cost Impacts	None	Cost increase for LEED-compliant roofing materials $100,000	1. Additional structural steel $20,000 2. Cost increase for green roofing $400,000 3. Less land needed for stormwater retention $200,000
Life-Cycle Benefits	None	Energy cost reduction $20,000/year	1. Energy cost reduction $20,000/year 2. Maintenance cost reduction $5000/year

to take the green roof option off the table just yet. A green roof also contributes to achieving LEED's credits for open space and habitat preservation, as well as providing a vital amenity in an urban environment, viewable for passive recreation or used for active recreation. As an example of passive recreation, the architecture firm Cook+Fox in New York City created a green roof on the eighth floor of a 100-year-old building in downtown Manhattan, as part of a LEED for Commercial Interiors Platinum tenant remodel. The roof is available for passive enjoyment by all employees and visitors. While it's not accessible to the occupants, it is visible through many of the firm's windows, providing a vital visual amenity in a heavily congested urban area.

WATER EFFICIENCY: CREDIT 1—WATER-EFFICIENT LANDSCAPING

Let's look at a very different example, addressing the concern for reducing potable water use for landscape irrigation. Two alternatives are to plant only native/drought-tolerant species or to collect rainwater and use it for irrigation. In the second option, we add $80,000 to costs for collecting, treating, and distributing collected rainwater and graywater to our irrigation system (Table 8.4). For the first option, we also save $15,000 per year in purchased water costs and $5000 in irrigation system maintenance, with immediate payback. The second option also yields $15,000 per year savings in water purchases. It requires more than 5 years to pay for itself. In the analysis, the use of native plants provides the economically superior solution and meets our Triple Bottom Line goals as well (Fig. 8.3).

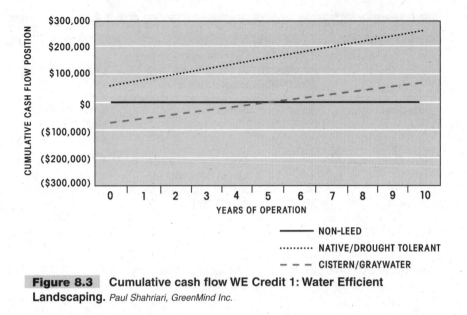

Figure 8.3 Cumulative cash flow WE Credit 1: Water Efficient Landscaping. *Paul Shahriari, GreenMind Inc.*

WATER EFFICIENCY: CREDIT 3—WATER USE REDUCTION

Let's look at another water-related situation. In this case, we want to reduce water consumption in the building, a key issue identified in our visioning session, not only to save operating costs but also to create environmental benefit and to reduce the impact of development on urban infrastructure. Our options include specifying

TABLE 8.4 WATER EFFICIENCY: CREDIT 1 WATER-EFFICIENT LANDSCAPING			
	NON-LEED BUILDING	**LEED CREDIT SOLUTION #1 NATIVE/DROUGHT TOLERANT PLANTS & NO PERMANENT IRRIGATION**	**LEED CREDIT SOLUTION #2 RAINWATER CISTERN & GRAYWATER SYSTEM**
Soft Cost Impact	None	None	Additional MEP engineering $5000
Hard Cost Impacts	None	1. Decrease in landscape material costs $20,000 2. Elimination of permanent irrigation system $25,000	1. Rainwater cistern $50,000 2. Graywater system $30,000
Life-Cycle Benefits	None	1. Water cost reduction $15,000/year 2. Irrigation system maintenance cost reduction $5000/year	Water cost reduction $15,000/year

TABLE 8.5	WATER EFFICIENCY: CREDIT 3 WATER USE REDUCTION		
	NON-LEED BUILDING	**LEED CREDIT SOLUTION #1 WATER-EFFICIENT FIXTURES**	**LEED CREDIT SOLUTION #2 RAINWATER CISTERN WITH FILTRATION**
Soft Cost Impacts	None	None	Additional MEP engineering cost $5000
Hard Cost Impacts	None	Low-flow fixture premium $5000	Rainwater cistern $25,000
Life-Cycle Benefits	None	Water cost reduction $5000/year	Water cost reduction $10,000/year

water-conserving fixtures or using captured (and treated) rainwater in place of purchased potable water for toilet flushing and urinals (Table 8.5). In many projects, of course, these approaches are used together, but for the purposes of illustration, they are considered separate ways to get to the same goal. You might also note that if we use both measures, such that the total water use is reduced by 50 percent, then we may also achieve another LEED credit point, Water Efficiency credit 2. If the building is a low-rise building with a large roof area, typical in many situations, we might have enough harvested rainwater to handle both the site irrigation and the internal water-use reduction goals.

The simple change-out of plumbing fixtures has about a 1-year payback, while the rainwater harvesting, treatment, and distribution system has a greater annual benefit, but a 3-year payback (Fig. 8.4). Note that all rainwater collection and reuse options

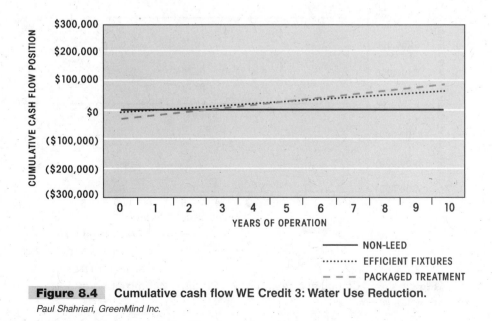

Figure 8.4 Cumulative cash flow WE Credit 3: Water Use Reduction.

Paul Shahriari, GreenMind Inc.

currently require some extra maintenance as well as dual-plumbing systems inside of buildings, a cost not incurred by the simple option of choosing lower-water-using fixtures. Over the short run, the efficient fixtures provide the better solution, but these two approaches are not mutually exclusive and both may remain on the table for further consideration.

ENERGY AND ATMOSPHERE: CREDIT 1—OPTIMIZE ENERGY PERFORMANCE

What about energy, which is typically the larger concern in most green building projects? How do energy conservation and renewable energy options stack up in the EVA analysis? Let's look at the first LEED credit for energy conservation and efficiency measures, worth from two to ten points. In the current LEED-NC 2.2 system, each new construction project must save at least 14 percent against the ASHRAE 90.1-2004 standard, to meet a prerequisite. Each building renovation must save at least 7 percent against the same standard to meet the prerequisite. The lower threshold results from the fact that existing building renovations typically do not change the insulation in the building walls or roof, even while they might replace older windows with more energy-conserving fenestration.

Energy efficiency is a big credit in most high-performance projects and no short analysis will do justice to the complexities involved. In this case, we look at upgrading the building envelope and lighting systems, as against improving the efficiency of the HVAC systems and getting daylight into the building (Table 8.6). Figure 8.5 shows how we can analyze these complex measures with the EVA tool in the *Ecologic3* project management software. The payback period for the envelope

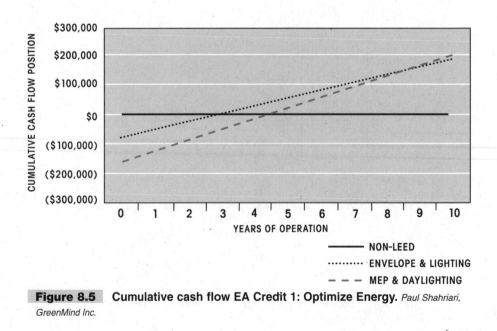

Figure 8.5   **Cumulative cash flow EA Credit 1: Optimize Energy.** *Paul Shahriari, GreenMind Inc.*

	NON-LEED BUILDING	**LEED CREDIT SOLUTION #1 BUILDING ENVELOPE & LIGHTING UPGRADES**	**LEED CREDIT SOLUTION #2: HVAC SYSTEM UPGRADES & DAYLIGHTING**
TABLE 8.6 ENERGY AND ATMOSPHERE: CREDIT 1 OPTIMIZE ENERGY PERFORMANCE			
Soft Cost Impacts	None	None	Additional daylight modeling $10,000
Hard Cost Impacts	None	1. Roofing upgrade LEED-compliant $100,000 2. Building envelope insulation upgrade $50,000 3. Reduced Heating & cooling equipment $75,000	1. Upgrade to water-cooled chiller $150,000 2. Add daylight-enhancing light shelves $150,000 3. Reduced lighting density $150,000
Life-Cycle Benefits	None	Energy cost reduction $25,000/year	Energy cost reduction $35,000/year

and lighting upgrade measures is about 3 years, while the increase in mechanical system efficiency, coupled with daylighting, results in a simple payback between 4 and 5 years. Beyond the 8-year mark, the HVAC and daylight package will be more cost effective, so it's worth considering for many buildings. In real life, these paybacks may be reduced further by using available federal and state tax credits and deductions, along with local utility incentives for conservation. Remember always to include incentive payments in all EVA analyses. (In the near future, one may also have to add a "carbon tax" to account for CO_2 emissions.) This example also illustrates another cardinal tenet of integrated design: always try your best to reduce energy demand (envelope measures) before trying to make the mechanical and electrical systems more efficient.

ENERGY AND ATMOSPHERE: CREDIT 2—ONSITE RENEWABLE ENERGY

Now let's take a look at installing renewable energy measures for our hypothetical project. In this case, we are almost always faced with considerably higher capital costs, relative to annual savings. However, an important mitigating factor may be the incentives available. Many states and utilities, along with the federal government, offer tax credits, and direct payments that can reduce the effective capital cost of such measures by 50 percent or more. Table 8.7 shows the EVA analysis for two options: a solar electric (photovoltaic) system or a wind power turbine, while Fig. 8.6 shows the cumulative cash flow from both project elements. Many projects are beginning to

TABLE 8.7   ENERGY & ATMOSPHERE: CREDIT 2 ONSITE RENEWABLE ENERGY

	NON-LEED BUILDING	LEED CREDIT SOLUTION #1: PV SOLAR ELECTRIC SYSTEM	LEED CREDIT SOLUTION #2: WIND TURBINE SYSTEM
Soft Cost Impacts	None	Additional MEP engineering $20,000	Additional MEP engineering $20,000
Hard Cost Impacts	None	1. Roofing upgrade for PV system $50,000 2. PV system $500,000 3. Tax incentives $150,000	1. Wind turbine system $ 250,000 2. Tax incentives $75,000
Life-Cycle Benefits	None	Energy cost reduction $20,000/year	Energy cost reduction $15,000/year

consider these options, so pay close attention to the analysis! In both cases, with strong federal, state, and local utility incentives, the paybacks are less than 15 years. Since solar and wind technologies are visible statements of a low-carbon future, a project might well decide to include one or both for that reason alone. Looked at another way, a 13-year payback is nearly a 7 percent annual return at current energy prices. What you effectively get is an inflation-protected bond yield that compares favorably with similar investments.

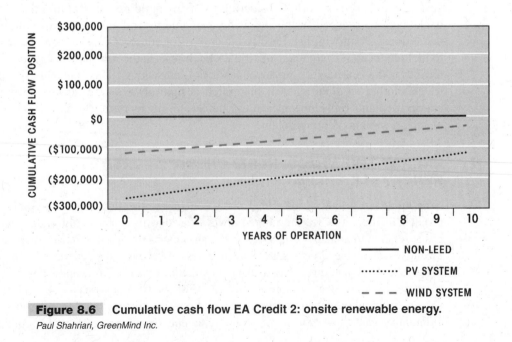

Figure 8.6   Cumulative cash flow EA Credit 2: onsite renewable energy.

Paul Shahriari, GreenMind Inc.

TABLE 8.8   INDOOR ENVIRONMENTAL QUALITY: CREDIT 3 CONSTRUCTION IAQ MANAGEMENT

	NON-LEED BUILDING	LEED CREDIT SOLUTION #1 BASIC IAQ MANAGEMENT PLAN	LEED CREDIT SOLUTION #2 ADVANCED IAQ MANAGEMENT PLAN
Soft Cost Impacts	None	None	None
Hard Cost Impacts	None	1. MEP equipment protection procedures and materials $13,000 2. MERV 8 filters $ 2000	1. MEP equipment protection procedures and materials $13,000 2. Temporary dehumidification/fresh air system for duration of finishes $100,000
Life-Cycle Benefits	None	1. Maintenance cost reduction $10,000/ year 2. Liability cost reduction $50,000/year	1. Maintenance cost reduction $20,000/ year 2. Liability cost reduction $100,000/year

INDOOR ENVIRONMENTAL QUALITY: CREDIT 3.1— CONSTRUCTION IAQ MANAGEMENT

Let's look at a less obvious situation, maintaining indoor air quality (IAQ) during construction. The choices are between basic and advanced IAQ management plans (Table 8.8). In this case, the advanced IAQ plan has a much stronger payoff during the 10-year life cycle, by reducing maintenance and insurance costs (Fig. 8.7). This case illustrates that it's worthwhile to search for both primary and secondary benefits,

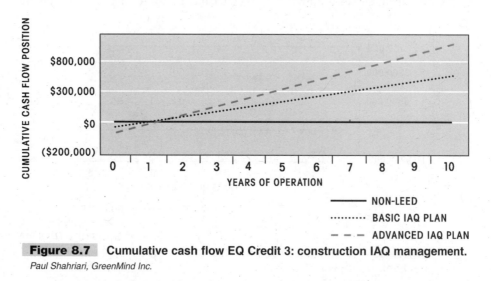

Figure 8.7   Cumulative cash flow EQ Credit 3: construction IAQ management.

Paul Shahriari, GreenMind Inc.

so that many of the LEED credits, which represent "best practices," are not excluded from consideration. These analyses help design teams evaluate alternatives that are in the LEED system but which may not be part of traditional design decisions. Because of risk mitigation and lowered maintenance costs, the advanced indoor air quality plan pays off in less than 2 years.

INDOOR ENVIRONMENTAL QUALITY: CREDIT 8.2—DAYLIGHT AND VIEWS

Let's look finally at two alternatives for daylighting and view creation (Table 8.9). In this case, the two alternatives, installing interior and exterior light shelves, as against providing north-facing skylights and reconfiguring the floor plate, have similar benefits (Fig. 8.8). Knowing this, we can turn to other design considerations because either solution will yield positive net benefits over the 10-year period of analysis.

SUMMARY OF THE EVA BENEFITS

During our review of all of the credit examples discussed above, we should always review how they impact our Triple Bottom Line EVA Log, as shown in Table 8.10. Not surprisingly, each of these measures creates strong triple-bottom-line benefits for each project. Creating high-performance projects is always going to be a balancing act between these benefits and their net costs.

ANALYSIS OF A COMPLETE PROJECT

When the team analyzes credits or solutions, there are many synergies that should be reviewed. Many credits and solutions on LEED projects are directly related to other elements within the project. These opportunities provide the team to really accelerate the benefit curves for projects and reduce the life-cycle paybacks of many items.

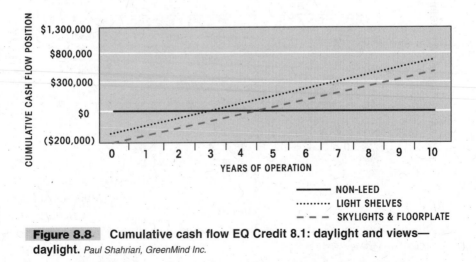

Figure 8.8 Cumulative cash flow EQ Credit 8.1: daylight and views— daylight. *Paul Shahriari, GreenMind Inc.*

TABLE 8.9 INDOOR ENVIRONMENTAL QUALITY: CREDIT 8 DAYLIGHT & VIEWS

	NON-LEED BUILDING	LEED CREDIT SOLUTION #1 INTERIOR AND EXTERIOR LIGHT SHELVES	LEED CREDIT SOLUTION #2 NORTH FACING SKYLIGHTS AND FLOOR PLATE RECONFIGURATION
Soft Cost Impacts	None	Additional daylight modeling $10,000	Additional daylight modeling $20,000
Hard Cost Impacts	None	1. Exterior daylighting shelves $80,000 2. Interior daylighting shelves $60,000 3. Reduced lighting density savings $40,000	1. North facing skylights and light wells $120,000 2. Floorplate reconfiguration with additional exterior skin $120,000 3. Reduced lighting density savings $60,000
Life-Cycle Benefits	None	1. Energy cost reduction $15,000/year 2. Productivity/retention/ absenteeism enhancement $100,000/year	1. Energy cost reduction $15,000/year 2. Productivity/ retention/absenteeism enhancement $100,000/year

TABLE 8.10 EVA BENEFITS

CATEGORY/CREDIT	PLANET	PROFIT	PEOPLE
SUSTAINABLE SITES			
Stormwater Design	Yes	Yes	Yes
Heat Island Effect	Yes	Yes	Yes
WATER EFFICIENCY			
Water Efficiency Landscaping	Yes	Yes	Yes
Water Use Reduction	Yes	Yes	Yes
ENERGY & ATMOSPHERE			
Optimize Energy Performance	Yes	Yes	Yes
On-Site Renewable Technologies	Yes	Yes	Yes
INDOOR ENVIRONMENTAL QUALITY			
Construction IAQ Management	Yes	Yes	Yes
Daylight & Views—Daylight	Yes	Yes	Yes

When the team finds credits that align with the owner's Triple Bottom Line Goals, then they can evaluate the level of certification to which the project will aspire. The life-cycle benefits of building green depend a great deal on the inflation of the operational cost factors. Operational costs such as energy, water, and maintenance are trending higher each year. Some operational costs are increasing fast, some as high as 10 percent or more per year. When analyzing the life-cycle benefits of building green, these factors should be taken into account.

We will now evaluate a complete example LEED project. We have evaluated all four levels of certification to produce this data. Soft and hard costs have been calculated just as in the previous examples. Also, life-cycle benefits were calculated for all four levels of certification, for an analysis period of 10 years. The impact of inflation (above nominal rates of 3 percent) on the benefits was also evaluated and is shown for a 0 percent (Fig. 8.9) and 5 percent rate (Fig. 8.10). These examples highlight the importance of factoring in the inflation rates for operational costs when evaluating green building options, since many green building measures involve adding costs upfront to receive a stream of benefits later (such as savings in energy and water costs).

In Table 8.11, you can see the net benefits increase at each higher level of certification. Ironically, the Silver and Gold certification levels yield a faster payoff than just plain Certified, breaking even in the second year. Even the Platinum certified building in this analysis breaks even in the third year, without assuming any inflation of energy or water costs, for example.

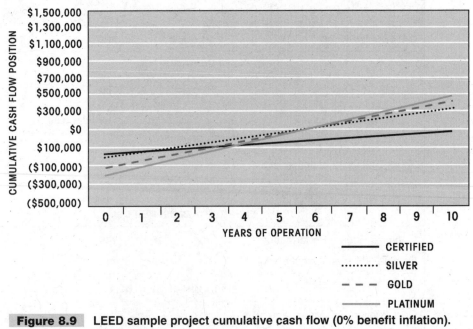

Figure 8.9   LEED sample project cumulative cash flow (0% benefit inflation).

Paul Shahriari, GreenMind Inc.

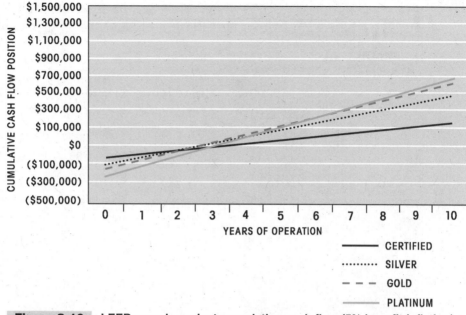

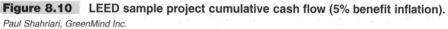

Figure 8.10 LEED sample project cumulative cash flow (5% benefit inflation).

Paul Shahriari, GreenMind Inc.

LENGTH OF ANALYSIS	CERTIFIED	SILVER	GOLD	PLATINUM
Year 0	$70,500.00	$100,500.00	$145,500.00	$215,500.00
Year 1	$36,250.00	$40,250.00	$72,250.00	$132.250.00
Year 2	$2,000.00	$20,000.00	$1,000.00	$49,000.00
Year 3	$32,250.00	$80,250.00	$74,250.00	$34,250.00
Year 4	$66,500.00	$140,500.00	$147,500.00	$117,500.00
Year 5	$100,750.00	$200,750.00	$220,750.00	$200,750.00
Year 6	$135,000.00	$261,000.00	$294,000.00	$284,000.00
Year 7	$169,250.00	$321,250.00	$367,250.00	$367,250.00
Year 8	$203,500.00	$381,500.00	$440,500.00	$450,500.00
Year 9	$237,750.00	$441,750.00	$513,750.00	$533,750.00
Year 10	$272,000.00	$502,000.00	$587,000.00	$617,000.00

TABLE 8.11 CUMULATIVE CASH FLOW—0% INFLATION ON BENEFITS (ENERGY COSTS, WATER COSTS, OPERATIONS AND MAINTENANCE, AND SO ON)

TABLE 8.12 CUMULATIVE CASH FLOW TABLE — 5% INFLATION ON BENEFITS (ENERGY COSTS, WATER COSTS, OPERATIONS AND MAINTENANCE, AND SO ON)

LENGTH OF ANALYSIS	CERTIFIED	SILVER	GOLD	PLATINUM
0	$70,500.00	$100,500.00	$145,500.00	$215,500.00
1	$36,250.00	$40,250.00	$72,250.00	$132,250.00
2	$287.50	$23,012.50	$4,662.50	$44,837.50
3	$37,473.13	$89,438.13	$85,420.83	$46,945.63
4	$77,121.78	$159,185.03	$170,216.66	$143,317.91
5	$118,752.87	$232,419.28	$259,252.49	$244,508.80
6	$162,465.51	$309,315.25	$352,740.11	$350,759.24
7	$208,363.79	$390,056.01	$450,902.12	$462,422.20
8	$256,556.98	$474,833.81	$553,972.23	$79,463.31
9	$307,159.83	$563,850.50	$662,195.84	$702,461.48
10	$360,292.82	$658,318.03	$775,830.63	$831,609.55

When we assume an increase of 5 percent annually on the benefits side, the breakeven years stay the same, but the Gold project, for example, has 30 percent greater benefits over the 10-year planning horizon (Table 8.12). For an institution or large corporation that has a long-term owner-operator perspective, assessing costs and benefits over a 10-year planning horizon is not unusual. When the benefits are known or can be estimated easily, then the issue is not "should we do this," but "how are we going to pay for it?" In other words, the decisions are financial and not economic in nature, because the economic value is quite clear.

Getting Started with Environmental Value-Added Analysis

Project teams should begin the green building process by identifying what Triple Bottom Line goals the project owner wants most. This ensures that the project team understands the owner's point of view for what will determine a successful project. Next the team should evaluate the various LEED credits and sustainable design solutions that support the owner's programmatic needs for the facility. Each item should be evaluated against the owners' EVA Log. The team should evaluate all impacts, examining soft and hard cost impacts and life-cycle benefits for each and every credit/solution on the project. The team should understand and review the net effects of all. The effects of inflation factors on life-cycle operations' costs play a vital role in evaluating solutions. The analyst should consider carefully the impact of even slight changes in operation cost inflation. The team can then assist the owner in selecting a set of solutions that fits their needs and aligns the best with triple bottom line goals established at the beginning of the process. By delivering the project to the owner in alignment with their sustainable goals and the programmatic needs of the project, the team can be confident of success. Most clients want to harness the economic value from environmental performance, and this approach helps you do that.

Ted van der Linden is director of sustainable construction at DPR Construction. He attests to the value of this approach:*

> There are not a lot of green building tools out there on the market today. We were an early user of Ecologic 3. We, at DPR, have created a lot of custom tools that are MS Excel macro-based. We have a tool called the Custom Delivery Model where we not only evaluate the "yes, no and maybe" for each of the different green strategies, but we also do a cost-benefit analysis of those strategies. For each credit you need, we have a yes, no or maybe and then we have another series of columns that list design-cost impacts, construction-cost impacts, and benefits for the first year, or payback if there is one. That way we have some quantifiable data on what the return on the initial investment is for the client. Ecologic 3 is a good tool for the pre-construction period, but it doesn't take a project all the way from start to finish. Without it, though, we would have been worse off. Once we've arrived at the precursor information—the LEED credits we are going to pursue, the first costs for those credits and the proposed benefits, Ecologic 3 allows you to input all of the information and then evaluate which credits have the most impactful return in the shortest period of time. You can select the strategies that make the most sense economically. It outputs, "Here's what LEED Silver would look like, here's what LEED Gold would look like, here's Platinum and here's just plain Certified." It shows the first-cost investments. If you look at the left side of column of a graph, it would show you that, for example, a Certified building has a first cost impact of $25,000, Silver has a first cost impact of $85,000, Gold has an impact of $1.1 million, Platinum is $1.6 million. It shows on a graduated scale when you've returned your first costs and when you're making money effectively. It's a great tool for pre-construction analysis. I think a lot of architects may not necessarily see the value because they don't have the answers to the information required by the software. But it's a great tool for people with the cost data.

Integrated Value Assessment

We interviewed Michaella Wittmann, sustainable design director at HDR, a large architecture-engineering firm with nearly 7000 employees. Her firm has developed a tool called Integrated Value Assessment that is used to guide high-performance projects.†

> As this industry is moving forward in sophistication, we strongly feel that it's critically important to work with building owners at the beginning of projects and help them

---

*Interview with Ted van der Linden, DPR Construction, February 2008.
†Interview with Michaella Wittmann, HDR, February 2008.

understand the value of sustainable design. We really try to use LEED as a tool for rating green buildings but prefer not to make LEED the focus of the overall goal. Rather, we talk about the owner's values and what is important for the building occupants.

We use a six-step integrated design process for sustainable design. Our process starts with setting client-specific and project-specific goals and prioritizing them. There's also a value piece where we try to figure out what's important to the client. The second step is identifying green measures. We categorize them in four different areas. They are: energy conservation and efficiency; indoor air and environmental quality; resource efficiency and effectiveness; and building occupants. In the past, clients really drove a heavy focus on the LEED checklist, and we found that didn't get us to the right solution. We've found that if we focus instead on these categories, it really helps us find the right solutions. After that, the third step is green building measurement. That might be LEED or another program or even our own metrics [another tool, discussed later]. The fourth is the selecting the solutions based on the measures identified in the second step. The fifth step is implementing them. The sixth step is called life-cycle management.

The thing that we've found critically important is the green-building measurement piece, which is our third step in the process. There's so much talk about life-cycle analysis and articulating the costs and life-cycle benefits of solutions to clients. We have a tool that looks at environmental and economic risk and benefits. We have internal group of economists at HDR who have worked with infrastructure and community projects. They apply fundamentals of good economic theory and process. But we've taken it a step further and look at environmental risk and benefit. We've come up with a set of inputs, some are specific to technology we're considering. Others are data such as the amount of carbon dioxide a particular technology might require or how much emission we might expect for a technology that uses X amount of fuel. We have industry-accepted data and we have data that's specific to the technologies and sustainable solutions we're considering. We're able to give our clients a pretty good idea what types of environmental risk and benefit they're going take on as part of a project.

For example, we used our economic and environmental assessment model on one of our Platinum Core and Shell office buildings. It was a spec office building and we had no way of measuring future productivity because we didn't know who was going to lease the space. We were able to take this model and say, "Given all of the sustainable solutions that we're using, such as a raised floor, photovoltaics, roof decks, operable windows, and all of the other items, what is the economic value as well as the environmental benefit of sustainable solutions we used?"

PLATINUM PROJECT PROFILE

McKinney Green Building, McKinney, Texas

Completed in April 2006, the McKinney Green Building is a 61,000-square-feet, three-story speculative office building. The facility is projected to reduce energy use by more than 70 percent and reduce water use by 30 percent compared to a similar, conventional office building. A rooftop photovoltaic system comprises 152 panels that provide approximately 10 percent of the building's electricity requirements. Two cisterns, each with a 9000-gallon capacity, collect rainwater for landscaping use. Nearly 18 miles of underground geothermal well piping and 120 wells in the parking lot will support the groundwater-based cooling system.*

We asked, "What if we estimated that productivity for the occupants in this building would increase by one-half of a percent because of the focus we gave on building occupant health?" We were able to model what the economic value might be for the client. That's been extremely valuable for us. We collectively look at the cost impacts of sustainable solutions and their benefit to the environment. We have printouts that show the expected reductions in carbon dioxide, for example, because of the solutions we're considering.

The tool is used for a number of different things. It helps clients understand the technologies. For developer-led projects, it helps them market the building. Our results showed that productivity far outweighs a lot of the benefits from even the energy side. You can show a client that a building might operate 40 percent more efficient in terms of energy, but if you can make building occupants one percent more productive, or if you can help with retention and recruitment, those benefits far outweigh the other benefits.

You can see that leading companies are all coming to the same conclusions. High-performance projects must include a strong analytical component that can be used not only to guide project decisions, but to convince skeptical owners that these measures have value in strictly economic terms. Part of the skill set of the integrated designer will have to include strong economic (as well as technical) analysis, as high-performance buildings move into the mainstream.

---

*HDR [online], http://www.hdrinc.com/13/38/1/default.aspx?projectID=300, accessed April 2006. Katie Sosnowchik, "McKinney Green Building Earns LEED Platinum Rating," iGreenBuild, June 5, 2007 [online], http://www.igreenbuild.com/cd_2876.aspx, accessed April 2008. Curt Parde, "What Makes the Building Green?," Environmental Design & Construction, November 1, 2006 [online], http://www.edcmag.com/CDA/Archives/506dd8b741fde010VgnVCM100000f932a8c0____, accessed April 2008.

9

GETTING STARTED—PREDESIGN
CONSIDERATIONS

Sustainable design projects often involve a myriad of considerations, ranging from seeking higher levels of energy efficiency to using recycled-content materials to incorporating daylighting, and so on. Many projects mistakenly assume that the only real question is whether to seek LEED certification from the U.S. Green Building Council. However, the decision to seek LEED certification, unsupported by a commitment to integrated design and to funding for the specific costs of certification documentation, is likely to be a recipe for frustration and ultimate futility. For this reason, many projects registered under the LEED system have failed to finish the process, as discussed in Chap. 5. According to those estimates, fewer than half and perhaps less than a third of all LEED projects registered through the end of 2004 had achieved certification by the end of 2007, even allowing for a three-year time lag between registration and certification.*

I support the LEED system and encourage my consulting clients to include it in their projects wherever possible; however, sustainable design involves a far broader set of considerations. The following sections incorporate and expand the LEED criteria to a broader range of design, construction, and operational considerations.

There are many excellent books and web sites that provide answers to specific sustainable design questions. However, often the issue is asking the proper questions at the proper time, to fulfill sustainability mandates from a client or owner. To draw out of building team members the full range of intelligence, experience, and expertise they already possess, I prepared a set of questions that apply just about any project with sustainability goals. These questions are organized by the phases of design and construction that naturally occur in most projects, and range from the general to the very particular, as the design process moves toward the final construction documents. You'll find them in this and the following chapters.

---

*"Where Are All the LEED Projects?", *Environmental Design & Construction*, July 2007, www.edcmag.com/Articles/Featured_Special_Sections/BNP_GUID_9-5-2006_A_10000000000000134921.

As someone once said, "ask a question, play the fool once; ask no question, stay the fool forever." Without asking the right questions at the right time, building teams take the risk of foreclosing good opportunities without even realizing it. The design and construction process has a pronounced bias against "revisiting" prior decisions; it always wants to move forward, unless forced to make significant system changes by value engineering requirements. Therefore, it's best to take a little more time at the outset of the project to ask good questions and to demand good answers. (Figure 3.2 shows how degrees of design freedom diminish throughout the process, and it's well known that the cost of making changes increases dramatically as the process of design and construction moves along. Most designers have seen this chart, but few truly appreciate the implications, in the rush to "pick up a pencil and start sketching.")

As the American inventor and design science guru Buckminster Fuller often wrote, a problem correctly stated is solved 100 percent of the time theoretically and 50 percent of the time in practice. By asking the right questions at the right time, you'll solve at least 50 percent of your design problem to implement appropriate solutions. I hope that this list of questions will lead design team members to better problem statements and then to improved sustainable design solutions.

Higher-Level Considerations: The Triple Bottom Line

Most of the LEED Platinum projects I looked at for this book had several similar characteristics. Many were institutional, they all had committed owners and they all hired very experienced design and construction teams, although for almost every team it was their first LEED Platinum project. John Pfeifer of McGough Construction in Minnesota talks about the importance of the owners' commitment and of early planning in their LEED Platinum project (Fig. 9.1), a headquarters for Great River Energy in Minneapolis.*

When you have a high-level LEED and high-performance building goal, [the importance of] all [early] decisions is incredibly intensified. You want to push decisions as early as possible because the potential downfall of mis-coordination and mis-communication is that the repercussions can be much greater, i.e., you might not get LEED Platinum or you might fail to receive the credits that you had planned on. Once you get past a certain point and you're committed to a certain building system, which maybe doesn't meet the requirement of your LEED credit scorecard, it may be too late to modify things. The difference is you need to expedite the decision-making process. You need to intensify the planning up front.

The piece that made all the difference in the work on the Great River Energy project was the fact that we truly had an owner that was committed and focused on obtaining these results. There are owners out there that talk about LEED and sustainability but as soon as the decisions get a little bit more challenging with respect to costs and energies that

---

*Interview with John Pfeifer, McGough, April 2008.

Figure 9.1 Aiming for a LEED Platinum certification, the Great River Energy headquarters in Maple Grove, Minnesota, expects to save more than $90,000 annually on electricity costs through the use of energy efficient measures and daylight harvesting. *Courtesy of Great River Energy.*

are needed to put towards solving problems and obtaining some of these goals, they may be not truly as committed as they once thought. Especially with a Platinum or high-level LEED-certified building, if the owner is not fully committed to the goals of the project, it won't happen. It cannot be left up to the rest of the project team.

There are a number of key questions that should be asked at the inception of every green building project, considerations that help define the project's key goals and "must have" project elements. Some of these include:

1 What are the project's specific goals and objectives (e.g., certification levels, meeting corporate or institutional policies, marketing to tenants or prospective buyers, creating extraordinary or landmark spaces, public relations, financial incentives)?

2 For institutional owners, the key question is: why is this building even needed? Are there existing facilities that would serve program needs that may have been overlooked? Can the building be smaller than originally planned? (For many green building advocates, the most sustainable new building is the one that's never built.)

3 Who are the key decision-makers setting the overall goals of this project and how will we be able to reach them throughout the process? (See the Yale case study for a classic example of how this was done well.)

4 What do we see as the key important environmental issues (in our region/to our company or organization) and how can we make a positive contribution toward addressing them with this project? (For example, water conservation in the southwestern U.S., habitat preservation, or restoration in most areas.)

5 To what degree do the project's financing sources (lenders and equity investors, for example) and other project stakeholders (public agency funders or university presidents or deans) hold the same philosophy with regard to making the project as green as possible?

6 What is the expected lifespan of the building (50 years, 100 years, 200 years)? How will this affect decisions relative to energy systems (e.g., passive vs. active, structural vs. mechanical), durability of materials selected, flexibility for future technology upgrades, allowance for changes in use (such as increases or decreases in occupancy), potential expansion of this building or nearby additions to the building stock, growth of vegetation to shade solar collectors, time-of-day use patterns, and flexibility issues?

7 What sources of financing and incentives (state/local/utility/other) are available for a sustainable project or for specific "green building" strategies? (I find that it's useful to look for outside money sources at a very early stage in most projects, to help pay for studies that aid decision making about green options.)

8 Who is responsible for investigating and securing these sources of financing? (Someone on the design team or owner's project team should be taking the lead in this effort.)

9 Are the stakeholders willing to spend the resources needed to secure required expertise (modeling/commissioning/additional design analysis/documentation) to implement high-level sustainability goals?

10 What is the minimum acceptable ROI or maximum "payback period" for costs that exceed the budgeted project costs? (Often, asking this question is very useful, since it alerts key decision makers of the need for looking for additional funds, e.g., for energy efficiency upgrades.)

11 What is the initial cost budget for the project? How and when (and by whom) was it determined? Are cost estimates still relevant; do they need to be revised to accommodate both inflation and green building goals? ("Where did this budget come from?" is a useful question, especially to expose assumptions and biases that may be hidden at the beginning of the project.)

12 Is the project cost realistic in light of construction cost inflation since the original budget was prepared? If not, are there alternative sources of funds, or does the project budget, scope, or scale need to be revised?

13 What economic, environmental, or cultural values does the client or owner ascribe to sustainable design? How will these be visible in this project?

One of the many significant high-performance projects finishing in 2008 is a new $270 million (Canadian), 690,000-square-feet, 22-story headquarters building for Manitoba Hydro, a provincially owned electric utility, in Winnipeg (Fig. 9.2). I spoke with both the owner's project manager and the architect and came away impressed by how hands-on the owner has been, without appearing to stifle the creativity of the design team. This project faced huge challenges, including an annual temperature

Figure 9.2 **The Manitoba Hydro office building in downtown Winnipeg is aiming for a LEED-NC Canada Gold or even Platinum certification.** © *Tom Arban Photography.*

swing of 70°C (126°F), not counting the wind chill in the winter when temperatures get to 40° below (astute readers will note that minus 40° is the temperature in both Centigrade and Fahrenheit)! Winnipeg is perhaps the windiest large city in North America, and it has cold, dry winters and warm, humid summers. To help the design team, the owner and architect called in one of the world's leading climate engineering firms, Transsolar of Stuttgart, Germany to help with the mechanical systems design philosophy and approaches. To understand the project better, I spoke with Tom Goldsborough of Manitoba Hydro, the client's project manager and Bruce Kuwabara, principal of KPMB Architects in Toronto, the design architect.*

The project began in 2002 with research into sustainable design projects, followed by an international design competition. The owner's stated goals were five-fold:

1 Demonstrate energy conservation, by saving 60 percent of the energy of a normal building built to Canada's Model National Energy Code; this goal was important since the owner is an electric utility looking to encourage others to save energy.

2 Create a productive and healthy work environment for Manitoba Hydro's 1800 employees, all of whom would be consolidated into a new headquarters building.

*Interviews with Tom Goldsborough and Bruce Kurabara.

3 Design a building with signature architecture that would serve the company's business needs for the next 50 years.

4 Help to revitalize downtown Winnipeg with a major corporate investment, by ensuring that the building design is open to the public and helps to create street life in the downtown area.

5 Make financial sense, so that it wouldn't have a large impact on electric rates.

Goldsborough and Manitoba Hydro's inside project team had one hidden strength: they really understood project management, because that's most of what they do at the company; build and operate large energy production and distribution facilities. So the prospect of managing a $270 million project was not that intimidating. To get started, the team hired an architectural consultant and visited 10 high-performance buildings in the United States and western Europe. After interviewing eight leading international architects, the Hydro team settled on KPMB, based on its body of work, experience with the Integrated Design Process and demonstrated commitment to sustainable design. (The architect of record is Smith Carter Architects of Toronto and Winnipeg.) Somewhat unique to this project is that all consultant contracts are with the owner, not the architect; the owner's goal was to be able to select the key individuals who would work on the project and to "incentivize right behavior." The project chose Poole Construction as the builder and Hanscomb as the Quantity Surveyor.

The project team held several multiday charrettes to tease out a project vision, delve into design issues and specifically to focus on obtaining a high level of energy efficiency in the building. The team hired an outside charrette facilitator and an energy expert. In the owner's view, this project had no important compromises: all of the original design goals are being met, and Goldsborough feels that the project could receive a LEED Gold rating, perhaps even Platinum, when all documentation is done. (As of May 2008, the project was close to the Platinum goal, according to the design team.) The delivery method was about 60 percent to 70 percent design/assist, and the owner thinks they might have even put too little focus on using the contractors to complete the design concepts. There is a postoccupancy evaluation process in place, and also a $1 million bonus to the design team if the project meets its stated energy performance goals.

KPMB's design work had a strong focus on Hydro's workforce and on creating a flexible building that could handle changing staffing needs and changing workplace habits over the next 50 years. The architect wanted to strongly link performance to aesthetics. Every design option was evaluated with three questions:

1 How does it work?
2 Does it work well?
3 How does it look and feel?

The climate engineer, Transsolar, made an early decision to decouple the ventilation system from the space conditioning system and to use the three, six-story atria in the building as a strong part of the fresh air and temperature regulation system. A system of 288 mylar cables in one corner of the south-facing atrium functions both as an art piece and for temperature and humidity control. In the winter, when the outside air is completely dry, water running down the cables humidifies the air in the atrium; in the summer, water is cooled to about 15°C and induces water vapor to condense on the cables, dehumidifying the air in the building.

Kuwabara has a few lessons learned that he shared with me for this book:

1 Innovation in a small market adds cost; in this project, there were not a lot of bidders to choose from.
2 The integrated design process must allow for serendipity, those "aha" moments. In Kurabara's words, the process must create "magic."
3 The most important issue is deciding exactly what to evaluate; in this project, the key issues were sustainability, connectivity of employees (to each other) and establishing a creative workplace.

He says, "it's important to keep your eye on the prize," and to evaluate against simple, clear goals and to do these evaluations in the right order, a message that is implicit in the organization of this book.

General Considerations: Sustainable Design

The LEED green building rating system was introduced in March of 2000, so this book is being written after the system has been in place for 8 years.* Still, most architects and building owners are embarking on their first LEED project, while others have completed only a few. Having accessible expertise can often mean the difference between success and failure. This section addresses these issues with the following questions.

1 Have we made a firm decision to follow an integrated design process, and are we making this process known to all potential proposers for design and construction services?
2 Have we adjusted our expectations and budgets to reflect the nature of the integrated design process?
3 Is this approach clearly stated in all our Requests for Qualifications sent out to the design and construction community?
4 Has a firm decision been made to seek LEED registration and eventual certification? If so, who will take the lead in providing the interface with the LEED process, the owner or the architect, or a green building consultant? (We're finding that many architectural firms want to manage this process in-house rather than

---

*LEED dates from March 2000, with the introduction of the LEED version 2.0 system for new construction.

defer to an outside consultant; however, the owner may want the expertise that a specialized LEED consultant brings. What's more important is that the responsible person has the authority to demand documentation, studies, and analyses on a timely basis from other building team members).

5 Are LEED certification requirements already contained in the scope of services for the design and construction team? Are extra costs for LEED analyses and actions built into the project budget?

6 If LEED certification is not yet a program element for this project, what will it take to secure this decision from the owner or client? If the owner's not willing to commit, can the building team push ahead any with making the project "LEED Certifiable?"

7 Have we selected design team participants and consultants who are experienced in sustainable design and construction? Are they willing to "push the envelope" in key design areas to help us reach our goals? (This is a critical decision, especially with the selection of the architect and mechanical engineer, the two key participants in most high-performance building projects; without an engineer willing to try new ideas and to be a part of the integrated design process, the LEED-certification effort is akin to pulling a tooth the old-fashioned way, with a string around the tooth tied to a doorknob.)

8 What sustainable design resources (financial, expertise, partnering, LEED project management software) among the stakeholders and design team members will be available for this project? Does the project team have the authority to use these resources, as it requires?

9 Have all key design team members taken sustainable design workshops, including the daylong LEED Technical Review workshop? If not, is some form of formal training in the LEED system part of the design team requirements? (After all, nearly 75,000 people have taken LEED trainings, as of May 2008; it's not unreasonable to ask people to "get up to speed" on the system if they want to work on the project.)

10 Are members of the design and construction team certified as LEED-Accredited Professionals? If not, will at least one design team member commit to becoming LEED Accredited during the project's early design phase?

11 Have we reviewed case studies of similar LEED-certified projects for inspiration and guidance?

12 Have we completed a review of the LEED Credit Interpretation Rulings (CIRs) for guidance as to how we might set up our project?

PLATINUM PROJECT PROFILE

Artists for Humanity Epi-Center, Boston, Massachusetts

The 23,500-square-feet, four-story Artists for Humanity Epi-Center houses art studios and gallery space in Boston. At $183 per square foot, the total project cost was $4.3 million. The energy cost for the building built to minimal ASHRAE

© Richard Mendelkorn, courtesy of Arrowstreet.

90.1-1999 energy standards would have been $3 per square foot ($68,000/year), the estimated energy cost for the Epi-Center is $0.56 per square foot ($12,732/year). A 45-kilowatt photovoltaic system was designed to produce 58,000 kWh of electricity per year, 156 percent of the building's electrical energy needs and 32 percent of its total energy needs. Ceiling fans, operable windows, and exhaust fans will be used for cooling since there is no refrigerant-based cooling system in the building, a measure that will reduce the total electricity use by 65 percent.*

Site Selection and Site Evaluation

Paul Meyer, F. Otto Haas Director of Morris Arboretum at the University of Pennsylvania, describes the sensitive issues of site selection and site design for a high-performance project:[†]

---

*U.S. Green Building Council [online], http://leedcasestudies.usgbc.org/overview.cfm?ProjectID=736, accessed April 2008.
[†]Interview with Paul Meyer, Morris Arboretum, March 2008.

The Morris Arboretum is in a stream valley known as the Wissahickon. This valley has a remarkable sense of place. We've been blessed with a lot of preserved open space thus far. Also nearby, historically there have been quarries of Wissahickon schist, which is a gray stone that has specks of mica that sparkle. There's just a special feeling to the valley in terms of the natural landforms, the architecture, the views, and the vistas. We just wanted to make sure that whatever we were building was related to that and respected the sense of place. But we didn't just want to build buildings that looked like 18th century buildings. We wanted something that was forward-looking but at the same time respectful of the past.

As an arboretum, we are very committed to doing buildings that function also as exhibits. For example, back in the mid-1980s, the Arboretum built a parking lot in the middle of the arboretum. It was a controversial thing—like the line from the old Joni Mitchell song: "They paved paradise and put up a parking lot." This time, we wanted to do that but in the most beautiful and sustainable way. We created a parking lot that not only serves as a functional parking lot but it also functions as an exhibit on how to design parking lots that are environmentally friendly. Similarly, with this building, we wanted to make sure that it reflected things that one can do to enhance sustainability. Not only were we doing those things but, where possible, we wanted those things to be visible and to interpret them to visitors. The building itself will function as an exhibit on sustainability. That was something that we communicated from the beginning to the architects. They weren't just creating a building but they were creating an exhibit.

PLATINUM PROJECT PROFILE

William A. Kerr Foundation Office, St. Louis, Missouri

Originally built as a bathhouse in 1895, 21 O'Fallon Street housed an auto body shop before the Kerr Foundation purchased it. The 4800-square-feet renovation project cost $1.5 million. In addition to serving as the office space for the foundation, the building welcomes local bicycle commuters by offering secure storage, showers, and changing areas. Onsite power is supplied by a wind turbine and photovoltaics, which supply 25 percent of the building's annual energy demand. An underfloor air distribution system ensures high indoor air quality, comfort, and reduced energy use for ventilation. Low-flow fixtures and dual-flush toilets reduce the building's potable water demand. Rain barrels, bioswales, and a green room contribute the management of stormwater. The project is also designed to catalyze revitalization of the blocks around it, currently in a run-down state.*

---

*Vertegy (August 13, 2007), "Vertegy Delivers St. Louis City's First LEED Platinum Building," http://www.vertegyconsultants.com/news.cfm?id=9, Press release, accessed April 2008.

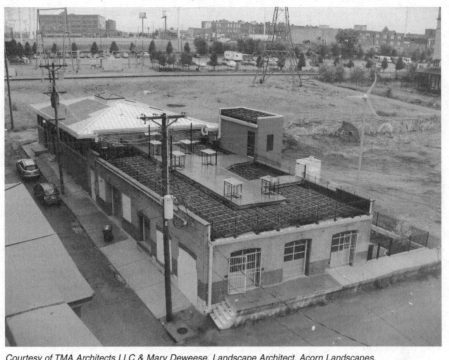

Courtesy of TMA Architects LLC & Mary Deweese, Landscape Architect, Acorn Landscapes.

From this project and many other examples, you can realize that site selection is often a critical issue in determining whether a project can reach high-level sustainability goals. The Sustainable Sites section of the LEED evaluation system offers considerable guidance for organizations that are flexible in their location criteria. Often we have no flexibility in site selection, for example, for a campus project or for an urban infill project. However, there are situations where two or more sites are being evaluated for a potential green building project. Key questions you should be asking during this phase of a building project often include the following:

1 Does the building program work better in an urban, suburban, or rural location? Can low-rise buildings be grouped for urban uses, in place of a high rise?

2 Who will be making the site selection decision? Are there explicit criteria? If we're going to use commercial brokers (to help us find a site), how will they understand what we're really after?

3 Are there restrictions on the sites under consideration that would hinder sustainable design in the project such as height restrictions, poor drainage or poor soils, a high groundwater table, or limited water availability?

4 Is the site located on prime agricultural land? Are there alternative sites that are nearby but are not on such a critical resource? (Note: in certain areas, such as California's Central Valley, without building upward, it's almost impossible to avoid using prime agricultural land for development.)

5 Is the site located more than 5 feet above the 100-year flood plain on a previously undeveloped site, to avoid both future flooding and impacts on neighboring buildings and land uses?

6 Is the site on land that provides habitat for a threatened or endangered plant or animal species? Have we performed a biological resources inventory of the site to determine its unique biotic characteristics?

7 Is the site within 100 feet of any wetland? Is there a possibility of using existing wetlands for stormwater management or wastewater management?

8 Is the site on current or previous public parkland (other than for parks projects)? If so, are there alternative sites that won't diminish the availability of parkland for a growing population?

9 Is there a need to complete environmental assessments or environmental impact reports on any of the sites under consideration?

10 Of the development sites we are considering, which are best served by existing or planned (and funded) public transit?

11 Which sites are within a quarter-mile of two or more bus lines or within a half-mile of a commuter rail, light rail, or subway station (existing or planned and funded)?

12 Are there a significant number of community services nearby, or will building occupants be compelled to drive for lunch, for errands, and for other frequently used services?

13 Is there an existing building available that would serve the needs of the project, which could be restored or reused? Are the costs comparable (for example, considering costs of required seismic or building code upgrades)?

14 Is the existing building on the National Register of Historic Places? Are sustainable design upgrades possible within the context of historic preservation?

15 What are the compromises with historic preservation, such as the inability to upgrade the energy performance of the building?

16 Are there local or state government tax or development incentives that would be important in deciding to reuse an existing building?

17 Are there "brownfield" sites that could be acquired and restored through the development of this project? Would this serve long-term public needs better than developing on a "greenfield" site?

18 Are there urban infill sites that would be appropriate for this project, in order to reduce infrastructure impacts and support existing transit and urban settlement patterns?

19 Do these infill sites contain a density of at least 60,000 square feet per acre (two-story or higher buildings) in the immediate vicinity of the project site, to encourage compact development?

20 Are there multiple uses that can be developed on the site that would help encourage transit use and the provision of community services (example: housing, retail, and office mixed-use)?

21 Are there existing natural site features such as trees, rivers, ponds, and wetlands that should be protected, maintained, or restored (such as waterways, canyons, or animal migration paths)?

22 Are there renewable energy sources (consistent strong winds/good solar orientation/ flowing water/geothermal) on or adjacent to the building site?

23 Are these resources available for use in the building project?

24 How can the site be best matched with renewable energy sources in the region?

25 Are there established neighborhood ecological patterns that could be enhanced, such as water flows and drainage patterns, local streams, green spaces, and/or transportation networks? Is there potential for contributing to the health of these systems? Can previously impacted streams or watercourses be restored as part of this project?

26 Can the building be sited closer to more of its users (following good planning dictates, "provide access by proximity first, then by transportation")?

27 Are there off-site areas that could be used as part of the project? (Example: a neighboring park or wetlands could be possibly used for occasional stormwater run-off.)

28 How might onsite or off-site features hinder the ability to make the project sustainable, particularly in terms of renewable energy (such as trees or planned future buildings nearby that could shade photovoltaic panels)?

29 Could a site be found that would make the utilization of renewable resources more visible? (In the case of the Hard Bargain Farm, profiled in Chap. 14, the project was broken up into two buildings, one in shade and one in sun.)

30 Can an adjacent building be used to support solar collectors to supply this project? (This is eminently possible with third-party solar investment partnerships that utilize federal and local incentives.)

31 Can an adjacent building be used for heliostats that might shine light onto dark parts of our site? (This might be a viable alternative for high-density urban sites.)

32 Are there adjacent buildings with dissimilar programs and time-of-day/week use patterns that would provide shared parking (examples: a church, a movie theater, a nearby shopping mall, and the like), so that we could reduce the amount of on-site parking?

33 Are there views that the building may take advantage of, or will the building obstruct the views of others?

34 What amenities near the building could be included in the building program (food services, fitness center with showers), without having to build them?

35 Is the building adjacent to sources of waste heat, graywater, reclaimed wastewater (purple lines), or other resources that could be positively utilized?

36 Are there opportunities to "harvest" rainwater from the site runoff for use in the building or for landscaping use?

> ## PLATINUM PROJECT PROFILE
>
> ### Audubon Center at Debs Park, Los Angeles, California
>
> Located 10 minutes east of downtown Los Angeles, the Audubon Center at Debs Park is a nature center within a 282-acre urban wilderness owned by the Los Angeles Department of Recreation and Parks. The total project cost (excluding the land) was $5.5 million. The design and construction of the 5020-square-foot building cost approximately $2.5 million, or $371 per square feet. Designed to use only 25,000 kWh of energy annually, the facility is operated completely off the grid; all of the power used is generated onsite. Estimated to use 70 percent less water than a similar conventional building, the center treats all wastewater on site. Fifty percent of the building materials was manufactured locally and 97 percent of the construction waste was recycled.*

Programming

In the programming phase, the amount of space and various uses for the individual users of the building are specified in enough detail for the designers to begin to place them within the physical structure. At this time, it's useful to pay attention to the energy and demand natural ventilation implications of space utilization in the building. Programming also specifies adjacencies, uses that need to be put close together. These decisions are often dictated by a client's organizational structure or by the nature of the work, such as a need to have researchers' offices and labs close together.

Let's look at the process for creating another LEED Platinum project, the Biodesign Institute at Arizona State University in Tempe, Arizona. There are two completed buildings adjacent to each other, one awarded LEED Gold and the other LEED Platinum. The design was a team effort between the nationally known lab design architect, Lord Aeck Sargent and the Phoenix-based Gould Evans. Jim Nicolow is a principal at Lord Aeck Sargent. He says:[†]

> What we've found on the Biodesign building (and with other projects) is if you start early with the intention of doing a green building, you can essentially get to Gold with a conventional project budget. It seems that Platinum tends to be that threshold where you really need to look at onsite renewable energy or technologies that have a higher cost. In the case of Biodesign in particular, it is two different projects, Building A and Building B. Building A is Gold certified and Building B is Platinum certified. The difference really was the PV array on the roof of Building B, which was enough to

*Building Green [online], http://www.buildinggreen.com/hpb/overview.cfm?projectId=234, accessed April 2008.
†Interview with Larry Lord and Jim Nicolow, Lord Aeck Sargent, March 2008; Trudi Hummel, John Dimmel, and Tamara Shroll, Gould Evans, March 2008.

nudge it over that threshold. We found that on other projects as well. With a conventional project budget, if you start out intending to build to LEED standards, Gold is achievable. It seems that to get to Platinum you have to do something beyond a conventional budget or beyond standard practice.

Also a principal at Lord Aeck Sargent, Larry Lord talks about how energy use is much higher in laboratory buildings and how it's possible to reduce that use dramatically with some clever design thinking and a full integration of architecture, engineering, and new technology:

Laboratory architecture and engineering is an interesting situation. It's in the nature of these buildings, (because of what's going on inside them) that you have to bring in air from the outside. In Arizona, in the summertime that air can be 110°F or more. You have to cool it down it to 55 degrees (for distribution within the building). So you have to put lots of chilled water through the coils get the air to cool down. Then you distribute it. When we first started the design we had [the standard] 10 to 12 air changes an hour, which means that 10 to 12 times every hour all of the air is new, so you can imagine how much energy that takes. Over time that has been reduced, and now we're down to four to six air changes an hour. The biggest challenge from a pure unadulterated energy usage standpoint is to try to reduce the number of air changes so we don't have to cool as much every time. Every laboratory building has the challenge of having a certain amount of air come in, stay there for 5 to 10 minutes (at the standard design rate of 6 to 12 air changes per hour) and then go out through the exhaust fans.

In terms of utilization, for example, the air is returned to the office side of the building. It's just a conventional office building but the extra air goes into the atrium and part of that goes into the labs. One strategy was, in essence, reusing the air that in an office building you would normally just re-circulate. We re-circulate a part of it but then it cascades down and is used as part of the air in the laboratories. We used all of the precautions about avoiding off-gassing; we flushed out the building before it got going, and so forth. You have to make sure that the beginning air quality is good.

In the labs, we know that the large number of air changes is going to give you very high quality air because just by turning over so many times that will clean the air. There's a new technique that's been placed in this building. In fact, we didn't even get a LEED point for it because it was installed later. It's a new system from Aircuity. Basically, it controls the air that's coming in by controlling valves that in turn control the number of air changes. Now that the system is installed, the number of air changes has been reduced to four per hour. The reason that's possible is that the system has, what I call, "sniffers" that sense if there's a problem. It can sense a number of gases and other contaminants. If it detects a problem, it kicks the air handler back up to 10 to 12 air changes per hour.

Nicolow says that the challenge for the design team (in lab projects) is to get the client's environmental health and safety people involved early, if you plan on reducing the air exchange rates. So there's another potential participant that needs to be a part of an integrated design process. In this project, Lord Aeck Sargent did most of the interior of the building and Gould Evans did the exterior shell. Lord says:

> We had set roles; having set roles on a project is really important. That made the relationship [between the two firms] work. It's just like in a theater production where you have this role and I have that role and you can't cross over roles, you have to do your thing in the production. It worked much like that.

The Gould Evans team of Trudi Hummel, principal in charge, John Dimmel and Tamara Shroll spoke about when they saw the integrated design process work, in very practical terms.

> [Hummel] It was a real team effort. In terms of the overall project, we tried to split it up evenly. So there was tremendous motivation on both of our parts to make it work. The issue of the interdisciplinary, interactive philosophy was really important in developing the overall concept for both buildings. The internal atrium is a very large, wonderful space, and natural light comes from the light monitors above. The discussion about how that would end up working was really a dual dialog. Often times there would be conflict, but in our minds that's not a bad thing. That's often what generates some of the best ideas.

A unique feature of this project is how the design team thought about water use and integrated the efforts of the landscape architect into the overall thinking.

> [Hummel] One of the project's significant features that really speaks of good sustainable design practices is the integration of the bioswale, which is a concept that was brought to the table by our landscape architect, Christie Ten Eyck. It's a feature that is really tied into the general workings of the building in the way that it captures the condensate water from the mechanical equipment. That water is stored and fed into the irrigation system. But also, it's an integral feature of the way that the landscape is a living place for the occupants of the building and visitors. It really is a big part of one's experience of the site. It has become one of the very well known features of the project, and it's something that we talk about a lot in terms of architecture and landscape working really well together.

> [Dimmel] To feed water into the bioswale, all of the roof drains had to be organized in way so that they all come down the side of the building and feed out to the bioswale. When it rains, it becomes an event out there. All of the rain from the roof is collected, brought down from the building, and redirected to the bioswale which is a low desert area. There was a lot of building system coordination to get all of those roof drains into that one area, crossing major utility lines and through the basement of the building. There was a lot of working that had to be done in order to get that to work but it came out to be something that was really spectacular and really speaks to what the research at the Biodesign Institute is about.

> [Shroll] There is a nice connection with the type of research that actually happens in the Biodesign Institute, which is so much about nature. What they're doing it is trying

to solve all kinds of problems in their research but they're doing it from a point of view of [how] nature [solves similar problems]. The idea was for them to have a connection from their offices into this amazing natural landscape that's all about the desert. It was a motivational and inspirational aspect of the connection between a building and its place.

The difficulties of making something happen is an example of how strong the project team has to be in terms of the strength of the design ideas and what the team is trying to achieve together. When there is such strength of ideas and such incredible support from the university's project manager, it makes some of the conflicts seem uneventful because everyone is working toward the same goal. Some questions to ask during the programming phase include:

1 How much building do we really need (more or less than first thought)? Can we afford what we need, or do we need to reduce the program? (This was done in the Ohlone College project, because the overriding goal was to have a high-performance green building, within a pre-established budget.)

2 Are our assumptions about building use still valid, with respect to emerging and established social trends, such as Internet use, telecommuting, 100 percent work from home, and so on? (In other words, do we still need all the space we thought we did?) If our assumptions are incorrect and we build too much space for our current needs, can we lease out space to others for 5 or 10 years?

3 Are there pieces of this building that do not require a permanent building to be constructed? For example, can seasonal and periodic uses be accommodated through temporary structures or partially enclosed spaces, for example, food service at a golf course? More architects are experimenting with temporary structures—in one case, a 10,000-square-feet LEED Silver sales office for a large Seattle development was specifically designed for deconstruction, so that it could easily be moved to another site when the development required the site it was on for future building phases.

4 Are there elements of our building program that can be overlapped in multi-use spaces that would reduce the size of the building? Shared parking is an obvious way to reduce environmental impacts, but shared meeting spaces in schools that can be used for community meetings at night are another obvious example.

5 Are we designing the building to be flexible enough to adapt to a new life after we've moved on or outgrown it? Can the interior walls easily be moved without affecting structural strength? (I visited a project in Oslo, Norway, a few years ago and found that it's quite common there for smaller buildings to be designed without any interior load-bearing walls, so that other uses can be more easily accommodated in the future, simply by moving walls.) New demountable wall systems are available on the market and help with this task.

6 Has the plan allowed for easy building expansion or expansion of cooling systems for future data centers or increased occupant load? It's easier to expect that we

might have to add more air-conditioning in the future than to design a building too rigidly for present uses. After all, "long life, loose fit, low energy" is one of the original definitions of sustainable design.

7 Should we build UP or OUT? There are cases to be made for both; a taller building tends to have less energy use per square foot of floor space, because there is less wall area and fewer roofs. A shorter building requires less energy use for elevators, and so on.

8 Are there FAR (Floor to Area Ratio) bonuses for pursuing green buildings or meeting other civic goals such as "1 percent for Art", which could affect the program of the building?

9 Are there places within the building that would engender a sense of community, such as gathering places, nature areas, winter gardens, or places of visual relief?

10 For a new project, can we size parking not to exceed local zoning requirements and provide preferred parking for carpools serving at least five percent of building occupants?

11 For a renovation project, can we add no new parking and provide preferred parking for carpools serving at least 5 percent of building occupants?

12 Can we provide alternative-fuel (electric, plug-in hybrid, or CNG) refueling stations for 3 percent of the total vehicle parking capacity of the site?

13 Can we provide a suitable means for providing secure bicycle storage for at least 5 percent of building occupants with convenient showers and changing facilities, at the rate of one shower per ten people?

14 Will everyone (or at least 90 percent of all occupied spaces) have a direct visual connection to the outdoors?

15 Are end-users involved in the planning to provide more accurate programming?

16 Can end-users be familiarized with building systems in advance of occupancy?

17 What degree of maintenance is available and what will be required for the building?

18 How much "churn" (i.e., annual turnover) is expected in office spaces, and how might this affect the choice of building systems and the economics of various building systems, such as underfloor air distribution systems?

19 Is this going to be flexible space, that is, open plan, and how might decisions about open versus closed spaces be examined for their impact on sustainable design options?

20 Can barriers to daylight and natural ventilation, such as high partitions and solid walls, be kept away from the perimeter to allow full daylight penetration?

21 Can the perimeter become a circulation zone, with private offices at the interior of the space, instead of the exterior, to facilitate daylighting and views of the outdoors?

22 Will the budget for this project support any additional design and/or capital costs needed to achieve the green goals?

23 Have we or should we set goals for onsite renewable energy (mostly solar) use in this project, say, of 2.5 percent of total energy use, 7.5 percent, 10 percent, or more?

24 Are there any additional budget requirements for sustainable elements available from internal or outside funding sources, such as third-party financing partnerships for solar photovoltaic systems?

PLATINUM PROJECT PROFILE

Center for Neighborhood Technology (CNT), Chicago, Illinois

LEED-certified in December 2005, the 13,800-square-feet building houses the Center for Neighborhood Technology, a Chicago-based nonprofit. With a cost of $82 per square-foot, this renovation cost considerably less than conventional rehabilitations in Chicago, which ranged at the time from $90 to $130 per square foot. The building was designed to use 50 percent less energy than a conventional building. A thermal storage tank serves as the building's primary cooling system and shifts peak electrical cooling loads to nighttime hours, when ice is made to serve the next day's cooling requirements. Recycled, regional, and healthy materials make up 13 percent of the building's total materials cost. Permeable paving was used for the parking lot, to encourage stormwater infiltration and the remaining site area is a rain garden.*

Predesign Work

One of the frequently overlooked aspects of LEED Platinum projects is that most of them had the contractor on board from the beginning, carrying out various aspects of predesign and preconstruction work. John Pfeifer is senior vice president at McGough Construction, Minneapolis; his team had just finished (spring 2008) a high-rise office project in that city which expects to get 58 points on the LEED scale and to receive the LEED Platinum rating. He speaks about the importance of the predesign effort:

> With this project in particular, we took those [collaborative] ideas to the extreme. We were involved with the pre-construction before the architect put any significant pen to paper. Early on in the project we had the entire team involved, not only the architect and McGough, but also a highly involved owner as well as all of the consultants. Because the Great River Energy headquarters wanted to become LEED Platinum, one of the first things we did was go through the LEED scorecard and understand all of the parameters, especially the parameters of the site location and geographical aspects and constraints. We established early on which points were easily obtained—if there is such a thing, which points we had no chance of obtaining, and then we analyzed everything in between. We rated the points by asking: Is it possible? To what extent? At what potential cost exposure? We then worked out that equation until we reached a point range of 55 to 58, which gives you a little comfort factor to hopefully achieve Platinum.

*Green Bean [online], http://greenbean.typepad.com/greenbean/2007/05/center_for_neig.html, accessed April 2008.

We would not have had a successful project had we not had the intensive early planning. The industry has proven over the years that any heightened cost of pre-planning—and this is probably true for any project whether it's high-performance, high-level LEED or not—if done efficiently and effectively, pays dividends in the end. If you can mitigate the need for last-minute changes, overtime, and an expedited schedule, then you will be way ahead of any additional cost spent in pre-planning. But it needs to be done efficiently and effectively. You can spin your wheels in pre-construction, too; then it can be money that can't be recovered and wasn't utilized very well.

This project illustrates the importance of asking the right questions during the pre-design phase of each project, to achieve high-performance results. Once again, we learn the truth of the old adage, "Don't just do something, sit there (and think)."

10

CONCEPTUAL AND SCHEMATIC

DESIGN

In the conceptual and schematic design phases, the high-performance design team typically investigates major systems alternatives for the project, looking at free natural resources such as solar, wind, and geothermal; climate control schemes; façade alternatives, building, massing, and orientation on the site. At this stage, the team can make some "back of the envelope" calculations of energy use and energy savings alternatives. The team can make a rough stab at costs, but doesn't know enough for detailed cost estimates. By the end of this phase, the team should know quite a bit about major building systems, pending detailed analyses in the design development phase, along with further costing and constructability reviews.

I cannot overstate the importance of considering sustainable strategies during these two phases; the entire reason for the questions posed in Chaps. 9 through 13 is to avoid overlooking good ideas in the rush to "do something." One authoritative text on schematic design puts it this way.*

> During conceptual design, the owner is convinced that the design team has a vision worth pursuing. During schematic design, the design team convinces itself that the vision sold to the owner is in fact feasible. Rarely do any big ideas creep into the design process after these initial phases.

In other words, every project starts out with a degree of trust, first on the part of the owner, that the team is up to the job. During the next phase, the team demonstrates that it can in fact implement the original project vision, or that it has to alter the vision owing to new discoveries: about the site, the stakeholder interests, the available resources, money, and so on. Once a specific design direction is chosen, almost nothing short of a major upheaval (such as a change in owners or owners' interests) will

---

*David Posada, in Alison G. Kwok and Walter T. Grondzik, *The Green Studio Handbook*, 2007, Amsterdam: Elsevier/Architectural Press, p. 18.

stop that train from leaving the station. Therefore, taking as much time as needed to think through all these questions is critical to setting the project direction on a sustainable track. The motto for sustainable design at this time might be "don't go charging ahead with designing, take time to think, study, analyze, and discuss)."

Conceptual and Process Questions

At this point in the project, the team should be comfortable enough with the integrated design process to begin answering and following questions.

1 To what degree do we want the building to be conspicuously "green," with environmental strategies on display, as opposed to having them operate quietly, behind the scenes?

2 Could the building educate others by making the green features more obvious, such as the use of photovoltaic solar panels or a green roof, or by opening up the internal workings of the building, to show systems such as an "enthalpy wheel" for energy recovery in operation?

3 What are the marketing benefits of such obvious green features to the client organization?

4 Is the entire team committed to an integrated design process? Is someone clearly in charge of the process? Have all team members signed a process document, committing themselves and their organizations to going forward as a group?

5 Have we considered a "sustainability forum" or "eco-charrette" to assist in conceptual and schematic design? Who will be invited to such a forum?

6 Have we considered a LEED-oriented design charrette to facilitate achievement of the green goals of the project?

7 If we're renovating an existing building, can we maintain at least 75 percent of the area of existing building shell (excluding window assemblies)?

8 For existing buildings, can we maintain at least 95 percent of the area of the existing building shell and at least 50 percent of the volume of nonshell portions (interior partitions, flooring, ceiling)?

PLATINUM PROJECT PROFILE

Chartwell School, Seaside, California

A private organization that helps children overcome learning disabilities, the Chartwell School serves children aged 6 to 14. Located in the California coastal climate zone of Monterey Bay, the Chartwell School followed the protocols set by the Collaborative for High Performance Schools, in addition to the LEED Platinum requirements. A 30-kW photovoltaic system produces 53,000 kilowatt-hours of electricity per year and offsets the production of 27 tons of CO_2 annually.

The project expects to reduce energy use 60 percent below that of a similar, conventional building. The design employs water-conserving fixtures, efficient landscaping and a rainwater cistern to reduce water use by an estimated 60 percent. Low-VOC finishes, natural ventilation, and CO_2 sensors contribute to high indoor air quality.*

Site Questions

Now it's time to answer specific site use questions in a more detailed manner. Often, a more detailed exploration of site opportunities and constraints can help inform the building design, in terms of choice of materials, building location on the site, and similar uses.

1 Are there materials onsite that can be used or reused? (Rocks; trees; clay for adobe bricks, soil for "soil-crete," rammed earth, soil "screened" for reuse, existing paving or concrete for fill or retaining walls.)

2 In what ways can we improve and/or limit our impact on wildlife habitat on and near the building site (example: can site design maintain or enhance wildlife corridors)?

3 Can we reduce the development footprint (including buildings, utilities, access, and parking), so that we exceed local open space requirements by 25 percent or more?

4 On previously developed sites, can we restore a minimum of 50 percent of the remaining open area by planting native or adapted vegetation?

5 On a new site, can we limit disturbance including earthwork and clearing of vegetation to 40 feet beyond the building perimeter? How will we write these requirements into the Division 1 specifications for the general contractor to follow?

6 Does the design work with the site or does it overly alter the site? (examples: extensive excavation/vegetation removal, disruption of habitat corridors, and the like)

7 If we have to alter some of the site, can we save the native vegetation and replant it onsite or possibly elsewhere, at a later time?

8 Is our program responding to the site's unique features such as topography, woods, pastures, or proximity to water bodies?

9 Does the site design create ecologically useful outdoor spaces; are we incorporating habitat preservation into the design, for example, by having slatted bridges over swales or streams so that sunlight can penetrate below?

10 Can water features be used for pedagogical purposes, for example, a pond for stormwater management or even constructed wetlands that can also be used by school classes?

---

*Chartwell School [online], http://www.chartwell.org/index.cfm?Page=132, accessed April 2008.

11 Can the circulation plan reduce the extent of impervious surfaces or find other ways to support vehicles, including emergency vehicles that require less paving?

12 Have we considered site restoration as part of the building program? If so, are we committed to creating natural areas versus providing active recreation areas?

13 Have we begun to consult with the landscape architect about site vegetation preservation and potential restoration, where appropriate? If this is an urban site, are we talking about saving water and creating habitat even in our hardscape plantings?

14 How can the design be made unique to the place and/or region, for example, through the use of regional or onsite building materials, or design references to local or natural features?

15 Are there landscape elements such as trees or watercourses that can be extended into the building, to connect indoors to out and thereby enhance the "sense of place"? What about "winter gardens" or other "inside/outside" features?

PLATINUM PROJECT PROFILE

Donald Bren School of Environment, University of California, Santa Barbara

The second LEED Platinum building ever certified, the Donald Bren School of Environment is an academic laboratory and classroom facility located near the Pacific Ocean in Goleta, California. The construction cost earlier in this decade for this 84,672-square-feet building was $26 million. Twenty-five percent of the building's energy needs are met by a combination of grid power from landfill methane gas and roof-mounted photovoltaic panels (which supply 7 percent of the building's energy). Energy savings are 49 percent compared to a standard ASHRAE 90.1-1999 building. Mechanical interlocks on the operable windows sense when the windows are open and automatically turn off the HVAC system. The project uses recycled materials throughout the building, including fly-ash in concrete (80 percent recycled), structural steel reinforcement (80 to 100 percent recycled), fireproofing material (made from gypsum, polystyrene, cellulose, and newsprint) and steel deck (30 percent).*

SITE WATER MANAGEMENT QUESTIONS

Water is emerging as a crucial design concern in many high-performance projects. An integrated design team will look at water in a much more holistic way, considering the entire "water balance" of the site. Some important questions include the following:

---

*Architectural Energy Corporation [online], http://www.archenergy.com/services/leed//donald_bren/, accessed April 2008.

1 How can stormwater be managed in the site design to reduce the impact of added impervious areas on storm sewers and waterways? Are we considering the use of bioswales and other methods for onsite retention of stormwater?

2 On a greenfield or newly developed site, can we implement a stormwater management plan that results in no net increase in the rate or quantity of stormwater runoff from existing to developed conditions?

3 On an existing developed site, with imperviousness greater than 50 percent, can we choose roofing, paving, and landscaping measures that will reduce stormwater runoff by 25 percent? What specific measures might we consider? Could this project incorporate a green roof as part of the stormwater management plan?

4 Have we studied the slopes and soils of this site to determine how best to manage stormwater onsite?

5 Are there soil-testing data available to determine percolation rates and the potential for groundwater recharge from natural rainfall?

6 Can we design and specify treatment systems for stormwater that will remove 80 percent of total suspended solids (TSS) by using Best Management Practices?

7 Have we consulted with the civil engineer and the local public works and planning officials about anything unique in our management of water resources on this site?

8 Will we be allowed to recover rainwater and reuse it onsite, or is it prohibited by state law, as is currently the case in Washington and Colorado?

9 Can the building project help to restore waterways running through, under, or adjacent to the site?

GREEN ROOFS AND LIGHT POLLUTION QUESTIONS

The subject of green roofs and light pollution reduction involves the architect, electrical engineer, and landscape architect and provides a forum for integrated design considerations.

1 Will rooftop mechanical equipment placement still allow for a green roof? What about smoke exhaust fans above an atrium? Can they be downsized to allow for more green-roof space?

2 Is a green roof feasible? Can the budget handle the additional cost? Will the stakeholders appreciate the amenity and practical values of a green roof? Will the green roof be available for tours or passive recreation?

3 Should the green roof be extensive (shallow) or intensive (deep)?

4 Will night lighting on the site not impinge on our neighbors?

5 Will night lighting (orientation and levels) on this site still allow for appropriate dark light levels suitable for nocturnal animals and birds?

6 Is there a "light pollution" ordinance in this jurisdiction that will affect nighttime lighting levels?

7 Can we meet the IESNA/ASHRAE light pollution requirement for the location where we're building or developing?

Water-Related Questions

Many people think that water will be the "oil" of the twenty-first century, the defining resource for civilization in many locations, energy use for water supply, and treatment accounts for more than 10 percent of building energy use. Therefore, questions about water are quite important to integrated building design.

1 Have we considered how to reduce, reuse, or recharge rainwater falling on the building? Do we have the budget for dual-plumbing systems, so that rainwater can be reused in the building?

2 Will the local jurisdiction allow graywater or rainwater to be reused for landscape irrigation and/or building water uses such as toilet flushing, cooling tower makeup water, parking lot washing, and so on? If not, can we appeal to their better instincts to begin allowing such systems?

3 Is there enough rainwater on the available roof area for catchment to supply public WCs and urinals for the building? Is the rainfall highly seasonal (as on the West Coast) or more evenly distributed throughout the year?

4 Is there adequate space within the building or on the site for underground or even surface storage of harvested rainwater?

5 Have we set explicit water conservation goals for this project, versus the 1992 Energy Policy Act plumbing fixture requirements (ultra-low-flow fixtures, water-free urinals, and the like)? Can we reduce water use by 30 percent or more compared with conventional buildings?

6 Are there local more stringent requirements for water conservation, owing to local climate or current drought conditions?

7 Have we established a baseline water use for this building that can be used to evaluate water conservation opportunities?

8 Can we reduce the use of potable water to flush toilets and urinals by at least 50 percent through the use of low-flow fixtures, dual-flush toilets, water-free urinals, and other means?

9 Have we considered the use of native and/or adapted vegetation for the site landscaping? Can we use plantings to create habitat for local wildlife?

10 Have we instructed the landscape architect to reduce water use for the irrigation of this site? By 50 percent? By 100 percent? Is everyone in agreement on a Xeriscape approach to landscaping?

11 If this is a high-rise building, can we use part of the fire sprinkler onsite storage tank also for rainwater storage, in this way combining uses and saving money?

PLATINUM PROJECT PROFILE

Philip Merrill Environmental Center, Annapolis, Maryland

The Philip Merrill Environmental Center in Annapolis, Maryland serves as the headquarters for the Chesapeake Bay Foundation. Completed in December of 2000, the total project cost for the 32,000-square-feet, 2 story building was $7.5 million. The project was certified as the first LEED Platinum project in the United States, under the original USGBC pilot program for the LEED rating system. The Merrill Center uses approximately two-thirds less energy than a typical office. The building's east-west orientation and operable windows were designed to take advantage of natural lighting and ventilation. The walls and roof were constructed with SIPs (Structural Insulated Panels). Approximately one-third of the building's energy comes from renewable resources including photovoltaic panels and geothermal heat pumps.*

Energy-Related Questions

In another context, the poet William Blake once wrote that "Energy is an eternal delight and he that desires, but acts not, breeds Pestilence."[†] This is a good introduction to the overriding issue today in high-performance building design, which is to realize our desire to dramatically reduce energy use in building design, construction, and operations. The pestilence in this case would be global warming from a continuing growth of global carbon dioxide emissions, nearly half of which come from commercial and residential structures.

More projects are striving to attain zero net energy use through the use of onsite power production and the purchase of green power to make up any shortfalls between onsite generation and actual use. Since most buildings rely on electricity for most of their power (using gas or fuel oil primarily for water heating and space heating), generating electricity onsite either through microturbines or solar photovoltaic systems is becoming an accepted way to deal with onsite power generation. Figure 10.1 shows the rationale for onsite production. The conventional remote electric power distribution

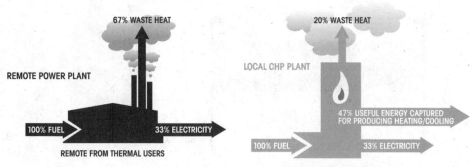

Figure 10.1 **Onsite energy production reduces the loss of value from the primary energy source by more than 70 percent.** *Redrawn with permission from Interface Engineering.*

*Chesapeake Bay Foundation [online], http://www.cbf.org/site/PageServer?pagename=about_merrillcenter_energy_main, accessed April 2008.
†William Blake, "The Marriage of Heaven and Hell," Plate 66–67, 1792, found in www.cyberpat.com/shirlsite/essays/blake.html, accessed April 28, 2008.

process loses about 67 percent to 75 percent of all primary energy on its way from the power plant to you; by contrast, onsite power production is about 80 percent efficient in delivering primary energy to end uses. Decisions about employing onsite energy production systems should be made during the schematic design phase.

At this stage of design, *how* we approach the problem is just as important as what we actually end up doing. A former colleague, Andy Frichtl, a mechanical engineer with Interface Engineering in Portland, Oregon, led the mechanical and electrical design team for the world's largest LEED Platinum building (as of the spring of 2008). He advances eight basic design approaches to energy-conscious design at this stage.*

- Estimate the amount and kind of free energy from onsite sun, wind, and water resources, along with seasonal air, groundwater, and earth temperatures.
- Using the building program, estimate daily, seasonal, and annual energy use patterns, including time of day variations.
- Estimate energy end-uses by type (heating, cooling, hot water, lighting, pumps and motors, ventilation fans, and plug loads) and then attack the largest end-uses aggressively (Fig. 10.2).
- Develop a plan for reducing end-use demand through building envelope strategies, higher-efficiency equipment, and daylighting, for example.
- Plan to harvest available natural resources.
- Plan to use energy storage systems to moderate peak loads and shift them to off-peak.
- Right-size all mechanical systems by using good analysis of actual requirements.
- Allow for easy expansion of mechanical and electrical systems to serve changing uses of the space, thus building flexibility into the sustainable design program.

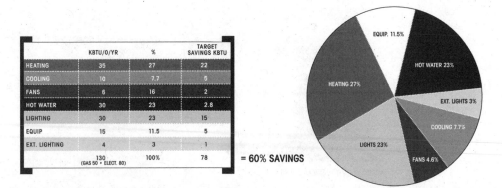

	KBTU/0/YR	%	TARGET SAVINGS KBTU
HEATING	35	27	22
COOLING	10	7.7	5
FANS	6	16	2
HOT WATER	30	23	2.8
LIGHTING	30	23	15
EQUIP	15	11.5	5
EXT. LIGHTING	4	3	1
130 (GAS 50 + ELECT. 80)	100%	78	= 60% SAVINGS

Figure 10.2 Early energy estimates for energy efficiency goals help teams to decide on the most productive areas for design explorations.

*Engineering a Sustainable World: Design Process and Engineering Innovations for the Center for Health and Healing at the Oregon Health & Science University River Campus, 2005, Portland, Oregon: Interface Engineering, Inc., p. 19.

Phil Beyl is an architect in Portland, Oregon and was the principal in charge of design for the world's largest LEED Platinum building. On the role of engineers at this stage in the design process, he says:*

We're in the camp that brings the engineers on board early [in the design process]. We think that's a pretty critical thing to do, in fact it's absolutely imperative. It may be different on client-by-client basis, because you don't want to barrage some clients with engineers who are speaking in some fairly technical language. But we typically work with a pretty sophisticated group of clients who are able to handle that.

What we expect out of our engineers is to be able to evaluate ideas on a conceptual basis. To able to use history from other projects and do some quick "back of the napkin" kind of calculations. For example, being able to say, "If you reduce the amount of glazing on the building 40 to 50 percent, that's going to result in a downsizing of your HVAC system by X tons and that's worth X dollars." It takes those types of quick evaluations to measure the quality of an idea. It may be a great idea, but it may only be worth pennies. It may be a really absurd idea, but it's worth lots of money [so it might be worth exploring further].

You always have to bring the idea back to a dollars and cents basis in order to advance it, because everybody has budgets. That's where several engineers, a lot that we work with, are really advancing their capabilities to think in more conceptual terms than in hard and fast engineering terms. Still, I criticize them often for not having enough people [on staff] who, in an engineering capacity, can think and communicate conceptually, as opposed to needing to know exactly how big every window is before they run a calculation for you. They're too accustomed to plugging these values into a fairly complex computer formula and letting the software run the results for them. They need to be able to pull back from that level of detail a little bit and give [the architects and owners] some higher-level guidance [especially at the early stages of design].

Here is a series of questions that you can use at this stage of the project to evaluate energy issues:

1 Can vegetation be placed on the south, southwest, and west sides of the building (particularly for low-rise and mid-rise structures) to reduce cooling loads in summer from the hot afternoon sun? In southern climates, have we also considered shading the east façade of the building?
2 Are we affecting solar microclimates or our neighbor's access to light and air, through the building height or mass, or even through landscape plantings, and what can we do to mitigate this impact?
3 How should we orient the building, and how will this impact our ability to utilize daylighting and passive solar design strategies? Are there site features or community connections that dictate how we orient the building? If we have a less than optimal orientation, owing to site constraints, how can we accommodate the building on

*Interview with Phil Beyl, GBD Architects, February 2008.

the site without using excessive amounts of energy or creating "hot spots" of over-heating?

4 Have we studied the potential for future vegetation growth around the building site, to determine if it might affect our design for daylighting, passive solar heating, and cooling requirements?

5 Is existing vegetation deciduous or evergreen, and can our design take advantage of those characteristics through passive solar features?

6 What elements should affect the building form, such as winds, sunlight, and topography?

7 Are we looking at higher ceilings, narrower floorplates and/or larger windows to support daylighting designs?

8 If this is a campus project or multiple-building project, have we considered arranging for a third-party to build, own, and operate a central chilled water or steam plant (to improve energy efficiency and reduce initial project cost)?

9 How much central operational control do we want over the building's environment? Who's going to be in charge of building operations, and how will they be trained?

10 What are the building's occupancy schedules and usage patterns? How likely are these to change in 5 years? Ten years? Longer?

11 What will be the major energy supply and use systems for this building? Can we specify and utilize energy systems that have better future potential cost control, such as geothermal heat pumps?

12 How many different ways can we think of to reduce energy demand in the building(s)? Have we started with ZERO as a goal, or are we just trying to reduce use from the level of a "code" building?

13 How much cooling will be required? Can we reduce this demand in creative ways, for example, by buying ENERGY STAR-rated laptops or flat panel monitors for everyone, to cut "plug" loads, or by requiring all tenants to use external power control devices for all electronics?

14 Is it possible to condition this space without mechanical cooling? How would that be done? Can we use "earth tubes" to precondition incoming air, or such innovations as chilled beams and passive downdraft cooling (in milder climates)?

15 Can we use generally more efficient (but higher initial cost) water-side (hydronic) heating and cooling systems for the project?

16 Will least-initial-cost considerations dictate instead of the use of air-side heating and cooling systems?

17 Can we design the building envelope thermal properties to eliminate perimeter heating systems, such as with triple-glazed windows in northern climates, or with newer window systems that can yield R-10 or better insulating values?

18 If geothermal energy is available in this region, have we considered its use for this project?

19 Have we considered the use of ground-source heat pumps for this project? Where would we put the pipes? Can they be laid out horizontally or must we drill vertical wells?

20 Is there a nearby source of cool water, such as a lake or ocean that could be used for cooling the building in summer and heating it in winter, using heat pumps (instead of "geothermal" can this be an example of "aquathermal" systems)?

21 Can our choice of structural and glazing systems influence the degree of infiltration from the outside and exfiltration of conditioned air from the inside, for example, by being easier to seal and keep sealed?

22 Have we begun to assess energy conservation opportunities for this project, considering not only building energy use, but also embodied energy of materials and transportation energy to and from the site?

23 Have we set energy savings goals versus state codes or versus ASHRAE 90.1-2004 (or later revised) standards?

24 Do we have the resources and ability in our consultant team to begin modeling energy use and natural ventilation options? Can we perform some early computational fluid dynamic studies of air movement inside the structure to assess potential for natural ventilation?

25 Will the local electrical utility or a local or state government agency pay for energy modeling or daylighting design efforts?

26 What envelope measures should we be considering for energy conservation? In addition, can we establish, for example, minimum SEER or EER levels for air-conditioning equipment?

27 Will our engineers specify variable frequency or variable speed drives (VFD or VSD) for all fans and variable air volume (VAV) controls for all mechanical air distribution?

RENEWABLE ENERGY QUESTIONS

Ultimately we will want to make up a certain percentage of projected building energy use with onsite renewables, especially solar electrics. How we go about thinking through this opportunity can often determine whether it's realized in the project.

1 Are we allowing for future solar installations on the building (in terms of designing roof surfaces and roof pitch)? Can we integrate photovoltaics into the south-facing shading of the building?

2 Are there ways we can assure of providing at least 5 to 10 percent (or more) of the building's energy use with onsite renewable (solar) energy?

3 Have we investigated local sources of "green power" for possible inclusion in the project? Is it possible to make this project "zero net energy," at considering site energy use?

4 Are green power sources available from the local electric utility, or will we be able to buy green power from a third-party provider? Are these programs "Green-E" certified by the Center for Resource Solutions or another acceptable independent third-party? Can we acquire Renewable Energy Credits (RECs) for this project?

5 What is the current and likely future premium for green power, and have we communicated this to the building owner or developer for consideration? Do we have policy guidance to buy RECs, or otherwise aim at a "zero carbon" building or facility?

6 If mechanical equipment will be placed on the roof, will there still be room for photovoltaic solar panels? Why can't the mechanical equipment be put in the basement instead?

7 Can we design our electrical systems to allow for future solar energy retrofits, by bringing wiring to the roof and allowing room for an appropriately sized inverter in the electrical room?

Paul Stoller is principal at Atelier Ten's New York office. This international firm is an expert in energy engineering and climate-responsive design. They were brought onto the Yale Sculpture Building and Gallery project by Kieran Timberlake Associates (see Chap. 3). Atelier Ten has completed a number of design projects at Yale, so you can assume they had the university's confidence in their abilities. Stoller spoke about his team's approach to this particular LEED Platinum project.*

> This was an odd project in that it was very fast track. Because of its fast-track nature, decisions had to get made quickly. It succeeded because decisions were made efficiently and quickly and because the design team was very skilled. We had good hunches about what would make for a high-performance building. We worked through those hunches in every project meeting, every two weeks when we had our regular session, and the design evolved very quickly.
>
> The conceptual design phase was the first half of the schematic design process. We did modeling in this phase, examining a whole series of major building performance options. We modeled the implications of glazing percentages on wall performance, of daylight performance, of control strategies, and of displacement ventilation versus mixed-mode ventilation. We looked at heat recovery ventilation versus no heat recovery. We looked at evaporative cooling versus no evaporative cooling. [In this way], we could quickly assess a long list of design options both architectural and mechanical.
>
> Then we looked at the interrelationships of those things. We always modeled things individually and then in combinations so we can see how they would affect each other. That happened in schematic design, and that's how we made our design decisions. That process used a schematic [energy performance] model so it was not a LEED-compliant whole building energy model [at that stage]. [We modeled] a somewhat abstracted form of the building [at this stage], but still one that represented how it would perform.

PLATINUM PROJECT PROFILE

Merry Lea ELC/Reith Village, Wolf Lake, Indiana

The Reith Village is an ecological field station for undergraduate environmental study for the Merry Lea Environmental Learning Center of Goshen College located south of Wolf Lake, Indiana. The Village includes two buildings for student housing and a third structure that serves as a learning center and environmental

---

*Interview with Paul Stoller, Atelier Ten, March 2008.

field station. Photovoltaic panels and a wind turbine generate more than 20 percent of the buildings' electrical needs. A solar thermal system supplies hot water for the occupants. Ground-source heat pumps reduce energy use by more than 60 percent compared to standard practices. Renewable and recycled products were used throughout the project. An underground cistern collects rainwater, which is filtered and pumped into the buildings for use in the laundry and for toilet flushing.*

Materials and Resource Questions

Efficient use of materials can be a consideration even in the early design stages. Building designers are increasingly asking for information on the life-cycle embodied energy and other environmental effects of structural materials choices.

1 Have we considered life-cycle assessment tools for analyzing key choices in terms of materials and energy/water systems for this building? Is anyone on the design team experienced in using these assessments, such as Athena or BEES?

2 If there is an existing concrete structure, what are the opportunities for recycling concrete from this site or nearby deconstruction sites? Can recycled concrete be used for aggregate or fill?

3 Are there regional sources for fly ash for use in the concrete mixture? Have we consulted with the structural engineers about using fly ash in concrete?

4 If this is a steel structure, does all the steel have documented high-recycled content?

5 Have the structural systems been assessed with regard to sustainability? Are we using lightweight concrete or less steel than normal?

Indoor Environmental Quality Questions

Since the major "business case" benefits of green buildings derive from gains in productivity and health, it's vital to start considering these issues early in the design effort.

1 If the building must have a large floorplate, can we design an atrium that will enhance daylight penetration and natural ventilation into all occupied areas of the structure?

2 Will the massing and orientation of the building support passive solar design and/or natural ventilation and daylighting strategies (Fig. 10.3)?

3 Do we have the consultants on our team who can model the effect of natural ventilation strategies? Are we committed to such formal modeling?

---

*Morrison Kattman Menze [online], http://www.mkmdesign.com/projects/sustainabledesign/h_1.htm, accessed April 2008.

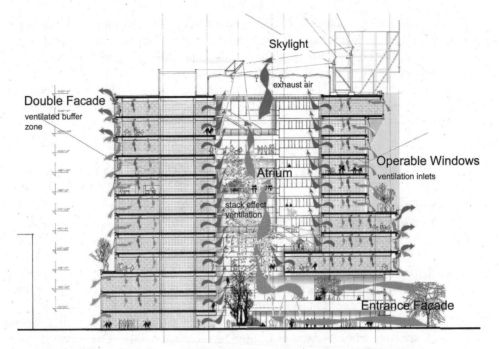

Figure 10.3 The Genzyme Center allows warm air to rise naturally through "stack effect" ventilation, to be exhausted through the top of the atrium. The exhaust air is replaced by cooler outside air most of the year. Operable windows and programmable blinds also help to moderate temperature swings and bring fresh air into the building. *Courtesy of Behnisch Architekten.*

4 Where do winter and summer winds come from? What are their frequency, magnitude, and duration? Are "wind rose" data from the site available from reliable sources? Do we have time to measure the wind resource for a year before the final design is set in stone?

5 How will localized wind directions and air pressures affect a design for natural ventilation? Will this information change as planned buildings are developed near our site?

6 Can we use internal circulation routes as air passages for natural ventilation? How can we help the building breathe? What will natural ventilation do to and for the building's interior layout?

7 Should the building be "sealed" for climate control, or can we open it up in various ways, with operable windows or "stack effect" ventilators?

8 Are operable windows compatible with other program needs? How will we communicate to building occupants when it's OK to open the windows?

9 Are we looking at radiant ceiling systems for heating and cooling? Are they accepted in this region? How will we control condensation and humidity?

10 Have we investigated using underfloor air distribution systems (raised access floors) for this project, both in terms of cost and technical feasibility?

11 If this building is in a northern (cold) climate, whether new or renovated, have we considered the use of "double-envelope" glazing systems, to allow for operable windows and natural ventilation inside the building?

12 Have we considered locating mechanical equipment near occupied spaces, to minimize duct runs? Or have we considered regrouping our spaces and adjacencies to minimize duct runs?

13 Does our design intent for indoor air quality specify meeting the minimum requirements of ASHRAE Standard 62-2007, to provide adequate ventilation?

14 Does our design intent for thermal comfort provide for compliance with ASHRAE Standard 55-2004?

15 If the climate is especially humid or extremely dry, do we intend to address these issues by installing permanent temperature and humidity monitoring systems with feedback to operator control?

16 Does our design intent for mechanical systems explicitly deal with the goal of having no exposure to environmental tobacco smoke, through separate ventilation of any designated smoking areas and through placement of air intakes away from places where people might be smoking outside the building?

11

DESIGN DEVELOPMENT

During this phase, we should have narrowed down our design choices to a few important alternatives that will help the building owner or developer realize the goals of the project. We need to complete the modeling of major systems and approaches and to give the design team clear direction. In some cases, we may decide, for example, to make heat recovery ventilation a major design focus (Fig. 11.1). We certainly need better cost estimates at this stage to make sure we are still within budget.

The public viewing window and computerized data display of the enthalpy energy recovery wheel earned the Newark Center for Health Sciences and Technology at Ohlone College an "innovation" point in the public education category, which contributed to its LEED-NC Platinum certification.

General Sustainable Design Questions

At this stage, it is often useful to take a step back and see where we are and whether earlier project goals need to be reexamined. We might also take a meeting or two to look at new opportunities that might have surfaced through the analytical efforts of the early design stages.

1 Has the project changed its green goals at this stage, and have we listed these in a way that all stakeholders are aware of them?

2 What are our goals for using life-cycle costing and life-cycle assessment methods for materials selection, and how can we use these and similar tools to help us decide on major building systems at this stage?

3 Have we evaluated all energy and water-using system alternatives as a team and do we fully understand the energy cycle and water balance for this project?

4 What are the costs of various strategies under consideration—compared to their environmental benefit? Compared to the increased productivity we can expect? Can we create a tool to examine these issues?

Figure 11.1   The public viewing window and computerized data display of the enthalpy energy recovery wheel earned the Newark Center for Health Sciences and Technology at Ohlone College an "innovation" point in the public education category, which contributed to its LEED-NC Platinum certification. *Photo courtesy of Lou Galiano, Alfa Tech Cambridge Group.*

Site Design Questions

During design development, we need to complete our site design and to ensure that we are maximizing opportunities for installing sustainability features through the project site.

1 Does the internal circulation of the project site support pedestrian and bike access? Have we designed separate entrances for bicycles, with power-assisted doors?

2 Are there adequate places for bicycle storage in the building? What about a location for an air compressor?

3 Are there showers located for easy access to bicycle entrance and storage?

4 If there is asphalt paving already on this site that will be removed, can it be reused for future paving?

5 Are stairwells and ramps centrally located, highly visible and easy to use, so that building users who choose not to use an elevator may easily move from one floor to another?

6 What is the light (pollution) impact of the building at night? Have we modeled the footcandle (fc) levels so that we know we can meet the IESNA 2004 (or current) site lighting standards in LEED?

Water Efficiency Questions

As we pointed out in the previous chapter, water conservation in buildings will undoubtedly assume even greater importance in the years ahead. As a result, during design development, we need to plan for a detailed systems approach to water efficiency.

1 Will ultra-water-conserving plumbing fixtures be acceptable to local officials (and possibly local plumbing unions)? Can we reuse gray water or other nonpotable water generated in the building?

2 Are there still other ways to reduce the water demand for landscaping, for example, by changing the aesthetic experience of the plantings?

3 Will landscaping provide shade for at least 50 percent of all impervious surfaces within 5 years, measured at solar noon at the summer solstice?

4 Will landscaping provide food and shelter for indigenous wildlife, by using native or adapted plantings?

5 If landscaping needs time for irrigation to get established, can we provide for a temporary (dry season) irrigation system that can be removed within 1 year?

6 If we need irrigation for this project, have we used high-efficiency irrigation controls to reduce such use by at least 50 percent over conventional means?

7 For the hardscape areas, have we provided high-albedo reflective surfaces such as light-colored concrete, so that there is less heat buildup in the local microclimate?

8 For parking lots, have we investigated the use of pervious paving systems (such as open-grid pavement systems) to reduce runoff? Would there be any time of maintenance that could reduce the effectiveness of pervious pavement, such as sanding of parking lots during winter?

9 Can we store rainwater or graywater cost-effectively on the site? Will the costs for site work for underground vaults or tanks increase significantly?

10 Have we discussed our water recycling or reuse systems with local code officials to get their feedback at an early stage?

11 If we're using rainwater storage, have we found a good location for the tank(s) or storage vaults?

12 Has equipment for treating harvested rainwater been sized and selected (possibilities include UV treatment, prefilters, piping connection to domestic water system)?

PLATINUM PROJECT PROFILE

The Ranch House at Del Sur, San Diego, California

The Ranch House at Del Sur is a 3000-square-feet welcome center for San Diego's Del Sur neighborhood. Completed in June of 2006, it serves as a community center and sales office. The Ranch House was designed to reduce energy use by 45 percent (excluding office equipment). A 5.7-kW roof-integrated solar electric system supplies more than one-third the facility's energy demand; overall energy savings are nearly 65 percent. Efficient plumbing fixtures have reduced water use by 40 percent. Reclaimed oak from a nineteenth century barn was used for the flooring, and timbers from an old pier were repurposed as ceiling trusses and trellises. The cabinetry and ceiling products were made from sunflower husks, and the insulation contains recycled denim.*

Photography by Reed Kaestner, courtesy of Black Mountain Ranch.

Energy Design Questions

Energy-related issues in the design development stage are so important to the future success of the project that it's worth spending a little time to see how it's done in high-performance projects. In the previous chapter, we quoted Paul Stoller about his approach to schematic design for Yale University's Sculpture Building and Gallery. Let's pick up the story in design development:†

Once we were in the design development phase, we did the traditional whole building energy model. We built and completed that before 50 percent design development (DD) so that we could validate it against our schematic design models. In the second half of DD we did testing on design options that were available at that time. At that

---

*Del Sur Living [online], http://www.delsurliving.com/ranchhouse.php, accessed April 2008.
†Interview with Paul Stoller, Atelier Ten, March 2008.

point we started looking at equipment selection and more basic control strategies. That level of design consideration comes up after design development, typically.

In the construction documents phase, we then updated the model a couple of times. Mostly at that time we were looking at refinements in the architecture—the type of glass, for example—or we were looking at control strategies in the HVAC. It was a standard process for us, but what was unique about it was how responsive the architects were. The design really moved a lot based on our input and the architects' critical review of this work with us. We were able to give more and better input than normal. Because they responded so quickly, we could do more and go further along in the design together in the amount of the time we had. It was truly an iterative process. [In the design development phase], we certainly looked at wall performance in great detail. There was a long discussion about the south façade, specifically whether it should be a double façade, which turned out to be prohibitively expensive (and we had some concerns about controls and reliability.) In the end, it became an externally shaded façade. Then Kieran Timberlake proposed backing up the spandrel glazing with Nanogel®-filled panels to make an exceptionally high-performance curtain wall. Kieran Timberlake used [Kalwall's] highest performing panel which is filled will an ultra-lightweight, high-translucent insulation, called Nanogel. It's exciting to use Nanogel because it performs so well, also it was interesting because the architects located the Kalwall inside the glazing, instead of its typical location as an external wall panel. Aside from the façade analysis work, we probably did more work on modeling HVAC on this project than we would do some other projects, to really confirm our hunches about the energy benefits of displacement ventilation and heat recovery ventilation.

Here are some key questions you can ask during design development phase, to make sure nothing has been overlooked that could significantly reduce energy use in the building.

1 Have we prepared hourly simulation models of the building design to assess energy-efficiency measures, before 50 percent design development?
2 Have we used this model to evaluate various envelope and system measures?
3 What is the size and location of all thermal mass in the project; can this mass be used as part of the heating and cooling properties of the building?
4 Can we use the project's natural ventilation goals as an approach to help us refine the sizing of windows and internal openings?
5 Can we train staff to operate and maintain the systems we are planning to use? If not, are there outside contractors who will be required to operate our facility?
6 Can the design take advantage of cooling strategies such as "night flushing" of the building or thermal energy storage to reduce mechanical cooling demand or shift its peak, thereby saving operating costs for utilities?
7 Are there innovative strategies for climate control, water system management and similar systems that are either economically feasible now or cost-effective over the life cycle of the building?
8 How will we document compliance with state energy codes applicable to this project? Will there be energy models integrated with our building information modeling (BIM) software?

9 If it's a small building (under 20,000 square feet), have we considered the use of the *ASHRAE Small Building Design Guide*'s prescriptive measures, instead of formal modeling?

10 Have we considered the use of an ENERGY STAR, LEED-compliant roof for this project? Have we included the energy benefit of this roof in our energy model?

11 If there is process energy use as part of this project, how will it be supplied and how can it be reduced?

12 If there is considerable use of thermal energy in this project, for example, for water or pool heating, can it be supplied by a cogeneration or microturbine system?

13 If there is a data center or other source of thermal energy built into this project, can the waste heat be recaptured for heating during winter for the building?

14 Have we investigated potential use of distributed energy systems such as fuel cells and gas-fired microturbines?

15 Can the emergency power system be designed so that it provides on-going power to the building for up to 72 hours, in case of an emergency?

16 How much can we reduce the energy load of the building and not decrease perceived comfort or occupant satisfaction, for example, by designing higher set points in summer and lower set points in winter?

17 If it was discussed in programming and schematic design, can our engineers assume a wider "comfort zone," perhaps 8°F (69°F to 77°F, for example), if we use natural ventilation strategies or energy savings approaches? Should we include operable windows or other means for occupants to control interior comfort conditions?

18 Will our tenants be willing to accept such expanded comfort zones (Fig. 11.2)? Can we write these expectations into the lease or incorporate them into tenant education programs?

19 Can we design the HVAC and fire protection systems to be free of HCFCs with global warming potential (GWP)? Or, can we take advantage of the tradeoff of GWP with ozone depletion potential (ODP) and still find acceptable HCFC refrigerants?

20 If there is base building refrigeration in the building (e.g., for food service, deli or restaurant), can we specify HCFC-free refrigeration equipment?

21 Can we recover waste heat from refrigeration equipment for water heating or other uses? (This is especially appropriate in grocery stores, restaurants, or cafeterias.)

22 Can the building incorporate thermal energy storage systems, either actively or passively, to reduce peak energy (typically cooling) demands?

23 Have we considered the use of an ice or chilled-water thermal energy storage system to reduce peak period electricity demand? Where will it be located? What are the current economics of peak demand reduction?

24 Have the expected future increases in future electricity pricing been considered in assessing the economics of various energy-using systems, for example, significant up-charges for peak-period power demand?

25 Have we considered the impact of possible future "time of day" or "real-time" energy pricing into our energy models and system choices? Have we provided submeters and a measurement and verification system that will allow us more information and control over energy end-uses?

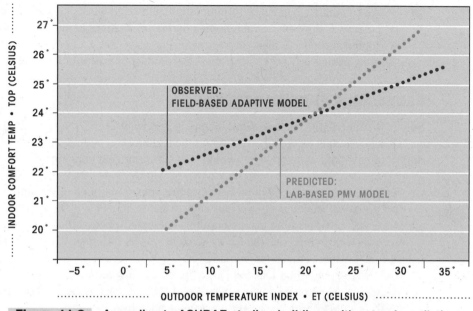

Figure 11.2   According to ASHRAE studies, buildings with natural ventilation have a wider range of acceptable temperatures for occupant comfort (from as low as 68 to as high as 80°F). *Data from University of California, Berkeley studies.*

26 What can an airflow (CFD) model tell us about sizing the building's windows and interior openings? What about the response of the building to varying wind conditions, both under existing conditions and in response to possible future location of buildings around this facility?

27 If we can ventilate the building naturally for some months of the year, how much energy would we potentially save? Do these savings justify the cost and potential discomfort of such approaches?

BUILDING COMMISSIONING QUESTIONS

Probably the single most important energy-saving measure for any building is commissioning, testing, and verification of all energy-using systems. Design development is a good time to hire a commissioning agent and incorporate that person or persons into the building team.

1 Has the owner decided whether the commissioning agent will come from the building team or will he or she be hired independently?

2 Has the commissioning agent or authority completed a focused review of design intent during or near the end of the design development stage?

3 Has the commissioning agent documented the owner's project requirements (OPR) and the basis of design (BOD)? Is there anything unclear about either?

4 Are we committed to enhanced commissioning throughout design and construction, and possibly continuous commissioning during occupancy? In this case, is the commissioning agent under contract to provide a focused review of building control and operating plans, early in the next stage, the production of construction documents?

RENEWABLE ENERGY SYSTEMS

Now is the time to get serious about including renewable energy systems in the project. These questions focus our attention on the technical, economic, and financial feasibility of solar and wind power especially, as strong project components.

1 How might the use of photovoltaic systems impact daily energy supply patterns and potential economics of various energy use systems?
2 Have we considered the use of building-integrated photovoltaics in the project? Because we can integrate the PV systems in several different ways, there needs to be time for these studies to be made.
3 If there are going to be spandrel panels in the curtain wall, for example, can photovoltaics be incorporated on the south face of the building? Are there economic benefits that would override the loss of power generation from having vertical solar panels?
4 Are there third-party partnerships that are willing to install a rooftop PV system at no cost to the project and simply sell us the electricity?
5 Can we use a nearby roof for the PV system and include it in this project's LEED certification?
6 Have we considered solar water heating for this project?
7 Have we considered a free-standing wind turbine as a signature design element for this project?

PLATINUM PROJECT PROFILE

Robert Redford Building, Santa Monica, California
The Robert Redford building provides offices for the National Resources Defense Council, a national, nonprofit organization dedicated to protecting public health and the environment. Other tenants include the David Family Environmental Action Center and the Leonardo DiCaprio e-Activism Zone. Completed in November 2003, the renovation of this 1917 three-story, 15,000-square-feet building cost $5.1 million. A 7.5-kW grid-connected photovoltaic system supplies approximately 20 percent of the building's electricity demand and contributes to a 55 percent reduction in utility costs (compared with a standard building). Treated rainwater and graywater are used for toilet flushing and irrigation. Dual-flush toilets and water-free urinals are used throughout the facility.*

---

*U.S. Green Building Council [online], http://leedcasestudies.usgbc.org/overview.cfm?ProjectID=236, accessed April 2008.

Craig Watts of MKK Consulting Engineers, Inc., in Denver, was the principal in charge for the LEED Platinum Signature Centre in Denver, for which Aardex, LLC, served as the owner, architect, builder, and developer, a very unique situation. He spoke about working with a multifaceted owner, evaluating technologies and making the compromises required for a successful project.

Working with owners who assume multiple roles on a project requires the ability to make distinctions. According to Watts:*

Sometimes it was confusing because we had to identify which hat they were wearing. Sometimes they were talking as the owner, other times as the general contractor. For example, they might say, "We have to do this or that because the owner needs to see the value." Then they would speak as the general contractor about timeframes for the project. Sometimes, the objective conflicted. As consultants, we had the opportunity to help them prioritize their goals.

Despite the confusion about the multiple functions of the project team being from the same company (if not actually being the same people), Watts says that there was a major benefit:

We didn't have to wait for decisions. It really streamlined the process because Aardex made the decisions as the owner, architect and contractor. It was truly integrated because all three roles were represented at the same time. Things aren't getting cheaper, and time is money. We didn't have to wait weeks to get decisions made.

The real lesson learned about this high-performance project relates to the role of the owner:

If you're going to do integrated projects, the owners can't be absent. They have to be involved in the design process. They have to be there making decisions with the design team to make it work right. That way they understand the whole process and they don't get a filtered version. They see the mechanical engineer, the electrical engineer, the architect and the general contractor all at the same time making decisions and the owners know how those choices impact each other. The integrated process is really about having everyone in the same room making decisions that benefit the project as a whole. If the owners are disengaged from the process, things get lost in translation. The project has to be the top priority; everybody else has to feel like they are winning in that process, but [in the end] it's all about compromises that make it work.

Compromises affect all aspects of the project including the section of technology:

For example, we used chilled beams in this project. They are installed around the perimeter of the building, next to windows. That's where the load is when the sun comes in [through the windows], and the chilled beams knock down the [heat from the] solar gain right there. The best thing for that [to be effective] is to use perforated

---

*Interview with Craig Watts, March 2008.

grills so there's more free area to make the chilled beam work. On the other hand, the architect needed to have a reflective surface to throw natural light deeper into the building. So we had to compromise between the airflow and reflective surface. We decided to use a grill that was open on one part and had a flat surface on the other part. It wasn't exactly what the architect wanted and it wasn't exactly what we wanted, but it really brings the light deeper into the space and it doesn't compromise the function of the chilled beams.

The project uses evaporative chillers but we could have gone even further with energy recovery [enthalpy] wheels [as did the Ohlone College project] that capture the heat or cool energy in the exhaust air from the building for energy and [use it to] pre-treat the air coming into the building. We did a quick calculation to determine how much efficiency we would get out of it. Was it going to get us from seven to nine LEED energy points? We calculated that it would get us close to eight maybe, and that it would cost another $60,000 to get to nine points. We quickly decided that we had done enough to achieve our goals.

The key to achieving LEED Platinum rested not only with the collaborative effort of the design and construction team, but also with many of the product suppliers:

People really went above and beyond as far as contributions and efforts (the suppliers and even the contractors). You know when you get into something and you don't know what it is? The Signature Centre did not start out on as a LEED Platinum project. It was determined to go for high-level certification after the venture was underway. Yet, everyone embraced the process. I think everyone believed in the project

Integrated design is really a collaborative effort. You can't have people sitting out on their own islands and being successful. The team has to have one single focus. If you have a player that sits over here and says, "I'm just going to do what I have to do and get out," then that's going to be a hindrance to the success of the project. Everyone has to have a commitment to the project not to just themselves and their bottom line. [This means that] you have to pick the right people. The selection of people is critical. They really have to understand how their work impacts other people. It's basically playing nice in the sandbox. It comes down to the relationships that you build.

This extended commentary about the role of all parties in the integrated design process illustrates clearly that everyone needs to be engaged and brought into the decision-making throughout the process. It also speaks to the need for rapid decision-making for high-performance projects; that in turn requires that the owner's project manager be capable of and authorized to make key decisions. For many large companies and institutions, that would be a departure from standard practice, in which the project manager does not have enough latitude to make key decisions. The lesson here is that high-performance results require process changes as much as advanced technologies.

LIGHTING DESIGN QUESTIONS

Lighting design is so central to the aesthetics of building design and to energy use that each project must "get it right" during design development. Fixture and lamp selection may be delayed into the construction documents phase, but the core lighting design must be completed in an integrated fashion during design development.

1 Have we reduced our design fc levels to 30 fc for most common spaces? Do we need to make special accommodations for visual comfort and for Boomers and older workers, such as task lighting in all workstations?

2 Have we investigated the use of direct-indirect parabolic trough lighting systems that can effectively provide ambient lighting with lower total energy use?

3 Have we looked at the use of T5 high-output lamps versus T8 lamps, to reduce fixture count and improve lighting levels? Have we specified compact fluorescents (CFLs) for all appropriate places?

4 Are we considering where we could use LED lighting, either now or in the future?

5 Are we considering the impact of interior color choices on design lighting levels and selection of luminaires, by looking at the reflectance of different color surfaces?

Materials and Resources Questions

Each sustainable design program has to place some emphasis on appropriate selection of materials. As with lighting design, we might leave some of the details until the next phase, but some basic choices need to be made at this time.

1 Does our design provide adequate space for recycling bins on each floor? Are we prepared to recycle, at a minimum, paper, cardboard, metal, glass, plastic, and batteries?

2 How will the building recycling system be integrated with areas for storage and collection of recycled materials for pickup and off-site treatment?

3 Can materials with recycled content be specified and purchased without compromising other needs such as performance, durability, and appearance?

4 Is each product we are considering recyclable after its use (e.g., "cradle to cradle" concept of materials use), such as various types of carpet?

5 Can salvaged or refurbished (better than recyclable) materials be used? Are there products that can be sourced from other locations, such as refurbished partitions or remilled lumber?

6 Do all the spaces need to have full finish materials, or can some be finished with the main structural materials (i.e., can the structure double as finish)? Have we considered the use of polished concrete, for example, as a finished floor, especially in transition areas?

7 If we are planning to use "structure as finish" to reduce the use of finishes in the building, is this acceptable to our client and the key stakeholders? If polished concrete will be used, have we taken acoustics into account?

8 What are trade-offs in using unconventional materials such as bio-based MDF board; can these materials have an instructional value through signage and other forms of communication?

9 Are the materials under consideration durable and easy to maintain, for example, flooring products and systems?

10 Where are the materials coming from—are they local or regional? Can we source them within 500 miles? How do we know we can do this?

11 Are life-cycle assessment data available for the materials we plan to use? How will we incorporate this information into our choices of materials?

12 What do we know about the environmental and business practices of the manufacturers? Are they acceptable to our stakeholders?

13 Are we designing for disassembly and recycling of materials during the life of the building or at the end of its useful life? What elements of the structure can be easily disassembled and reused, particularly if the site is needed for other uses? (This is easier in smaller buildings, of course.)

14 Does the design take advantage of standard dimensions for materials and assemblies, instead of specifying custom dimensions that might create more waste?

15 What materials entail the most toxic production processes? Are there workable alternatives to them?

16 What materials can help reduce or eliminate any toxic off-gassing in the air that our staff and clients will breathe? Can we avoid a lot of finished materials that have toxic or noxious chemical constituents?

17 Will the manufacturer reclaim materials (such as carpets) when they are ultimately removed from the project? What is the proof of that intention and/or practice?

18 If we are planning on significant use of salvaged, reclaimed, or recycled materials in the building, is this acceptable to our client and the key stakeholders? Will they see these materials as adding value or reducing it?

19 Is it possible to lease some of the building materials (such as carpet tiles) from the manufacturer rather than buying them?

20 Have we considered the use of carpet tiles instead of rolled goods, so that worn areas can be more easily replaced without removing good carpet?

21 What is the total embodied energy of this building (i.e., the energy required to extract, process, and transport the materials to the site and to build the building)? How can we determine this, or at least get a handle on it? Will it make a difference in our choices of finishes or other building materials?

22 Where did the wood for this project originate? Was the forest managed in an ecologically responsible way? Is a reliable third-party certification available?

23 Will the recycled products that we're considering hold up in the long run? Have we considered durability as a key component of sustainable design?

24 Will our design allow us to change out materials and whole systems in this building over its long life without major disruption? (This might dictate placing HVAC equipment in a more accessible location, for example.)

PLATINUM PROJECT PROFILE

StopWaste.org, Oakland, California

The 14,000-square-feet building serves as the headquarters for StopWaste.Org, the Alameda County Waste Management Authority and the Alameda County Source Reduction and Recycling Board operating as one public agency. Designed to be 40 percent more efficient than a typical office building, a few of the sustainable features include a photovoltaic panel array; efficient heating, cooling, lighting, and energy management systems; recycled and salvaged materials; and low-VOC paints.*

Indoor Environmental Quality Questions

Earlier, I pointed out how productivity and health concerns were tied up with employee satisfaction with the working environment. Studies reveal that indoor air quality and daylighting are two of the vital elements to introduce into any high-performance building project. Incorporating key indoor environmental quality features should be carefully considered during the design development phase.

1 How much could our staff productivity increase with better daylighting? With natural ventilation or 30 percent or more increased outside air? Have we accounted for the benefits of higher productivity in our cost-benefit model?

2 How could a daylighting model help us reduce energy use by optimizing the electric lighting design?

3 Are we considering the use of lighting controls that will allow a seamless blend of daylighting (when available) and electric lighting?

4 Will at least 90 percent of the regularly occupied spaces have direct views to the outdoors, through vision glazing (30 to 90 inches above the floor)?

5 Will there be a minimum 2 percent daylight factor for all regularly occupied spaces or for at least 75 percent of such spaces?

6 Have we considered reducing the organization's standard partition heights to 42 to 48 inches to accommodate views to the outdoors for everyone? How will acoustic privacy be handled in that situation?

---

*Rumsey Engineers [online], http://www.rumseyengineers.com/green_featured_stopwaste.php, accessed April 2008.

7 How will daylighting and natural ventilation be impacted by the choice of workspace partitions and their location? Are the interior designers fully incorporated into the project team?

8 How do we plan to control sunlight entering the building? Can we get daylight into the building without glare that would inhibit productivity?

9 Will we use fixed exterior shading devices, either horizontal or vertical?

10 Can we consider dynamic façade systems respond to changing solar orientation and exterior lighting levels?

11 Have we studied the use of light shelves for daylighting? How will they impact the building's aesthetics?

12 Have we considered the use of skylights and light wells to get light into the building? How will we deal with potential glare from such systems? Can they be adequately controlled in more southerly climates?

13 If the building is naturally ventilated, have we demonstrated with a CFD model that air distribution will provide laminar flow in at least 90 percent of all occupied spaces, for at least 95 percent of the hours of occupancy, so that everyone can share in the benefits?

14 Have we considered the use of operable windows in the building and located such windows? Is there at least one operable window for each 200 square feet of perimeter space (within 20 feet of the window)?

15 Are we going to provide for individual light, air, and temperature controls for at least 50 percent building occupants (e.g., through underfloor air distribution systems)?

16 If we use underfloor air systems, have we considered how these systems will be installed so they don't leak (air) and who will maintain the area under the finished floor to keep it clean?

17 Does our design ensure that ground-level or rooftop air intakes will not be impacted by sources of pollution, including HVAC exhaust, chemical uses, truck loading, smoking areas, and possible off-site pollutants?

18 Have we specified a permanent carbon dioxide monitoring system that informs operational adjustments of the ventilation system, so that we can have demand-controlled ventilation, especially in spaces with widely varying occupancy? Are we meeting ASHRAE 62.1-2004 standards for outside air delivery?

PLATINUM PROJECT PROFILE

Sweetwater Creek State Park Visitor Center, Lithia Springs, Georgia

Designed by the Gerding Collaborative, the 8700-square-feet Sweetwater Creek State Park Visitor Center provides exhibit space, offices, retail area, learning laboratory, classrooms, and restrooms. The project was completed in 2006 and cost $175 per square foot. Thirty-eight percent (nearly 4000 square feet) of the building's roof area is used for rainwater collection, to supply an estimated 44 percent

Courtesy of the Gerding Collaborative.

of the building's water needs. Along with low-flow fixtures and onsite wastewater treatment, rainwater harvesting reduces potable water consumption by 77 percent. Facility restrooms use composting toilets and water-free urinals. A 10.5-kW photovoltaic system produces approximately 20 percent of the building's electricity needs.*

---

*"Sweetwater Creek," Architecture Week, May 23, 2007 [online], http://www.architectureweek.com/2007/0523/ environment_3-1.html, accessed April 2008.

12

CONSTRUCTION DOCUMENTS PHASE

During the construction documents phase, it's time to provide the final details of design at a level that enables contractors to bid and build the project. During this phase, product and design specifications become "locked in" and are not easily altered. Therefore, decisions during this phase are important in determining whether we can reach the project's sustainability goals, especially if we're committed to using new methods or systems with which a local code official or contractor may not be familiar.

In this phase, we'll have a more intense focus on materials and resources. We'll consider using salvaged or reclaimed materials, perhaps from nearby demolition sites; recycled-content materials, such as fly ash in concrete or countertops made from recycled newsprint; locally sourced materials (instead of, for example, marble quarried in Italy, shipped to India for finishing, then shipped to the United States); bio-based materials such as bamboo, cork, linoleum, wheatboard cabinets, and natural fiber carpets; and sustainably harvested wood.

Energy-Using Systems

We also have to get very specific about the energy-using and energy-supply systems for the project. This activity is especially important as projects move toward "zero net energy" goals. A good example is an enthalpy wheel (Fig. 12.1) designed by Alfa Tech Cambridge Group of San Francisco, for the Ohlone College Newark Center for Sciences and Technology in Newark, California. Principal in charge Michael Lucas spoke about how this design element evolved.*

> We presented a number of mechanical concepts. One, of course, was photovoltaic panels to offset some of the electricity consumption in the building. We also presented three or four mechanical system options. *The two that turned out to be most the energy efficient and the ones that the client liked the most were the geothermal system and the enthalpy wheel energy recovery system.* (We also evaluated thermal storage and ice storage.)

*Interview with Michael Lucas, Alfa Tech Cambridge Group, March 2008.

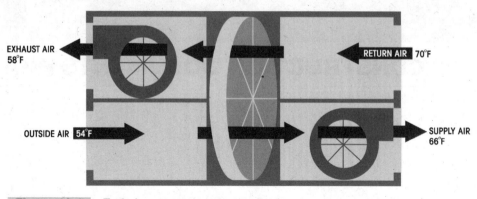

Figure 12.1 Enthalpy energy recovery wheel.

We did a number of computer models of each system that we were considering. Those were based on the building program that we had at the time, which initially was about 120,000 square feet. We modeled the building and then we did "what if'" scenarios for each of the systems we were considering. Then we did a life-cycle cost analysis, a performance analysis and a LEED point analysis based on the number of points each system would give us.

The decision was arrived at from several different considerations. One was, of course, cost. The second was what we would get in terms of reduced operational costs for the building. The third was what we would get in terms of the number of points from the LEED perspective.

One of the considerations early on was that the faculty wanted us to consider having operable windows. There's a downside with operable windows: they allow in pollutants, pollen, and dirt in some cases. Of course you can also have wind and other weather [to deal with]. But from an energy and engineering perspective, if you run air-conditioning or heating and you've got the windows open, then of course, you're consuming more energy, not less. As a compromise, since they wanted to have large volumes of fresh air in the building, we proposed the enthalpy wheel [which takes heat or cold from the exhaust air, depending on the season, and transfers most of it to the incoming air, thus saving most of the energy to condition the supply air].

Because there's little energy penalty for using this device, we almost tripled the amount of outside air in the building. Instead of having the normal, minimum code-required amount of air—and this is what got us another LEED point and helped us toward Platinum—we actually were able to provide a lot more air into the building without having operable windows. That air doesn't contain pollutants from the outside because it's filtered. When you go into the building, you get the sense that the windows are open but they're actually closed—they're not operable. You have a lot fresh air that adds tremendously to the learning environment. The faculty likes it, and the students like it even better, as it helps eliminate the complaint of wanting to snooze after lunch.

The enthalpy wheel can be very difficult to implement which is why you don't see these devices very often. They're huge. We have two of them in this project, and they're each 16 feet in diameter. They take up a lot of space and have a lot of associated ductwork. Architects, generally speaking, don't like to incorporate them in their buildings because they take up so much room. The architect, Perkins+Will, did a great job, making circular windows, like portholes, on two sides of the building so you can actually look through the windows and see the wheels rotating. Also there is a graphic display next to the window that shows how much energy is being saved real-time.

Because all of the air coming in and all of the air going out has to go through the system, it's got to be in a fairly central location unless you want miles of ductwork. The engineer has to work very closely with the architect to make sure you can locate them somewhere central in the building. It's perhaps 80 percent engineering and 20 percent architecture involved in getting these integrated correctly into the building.

A High-Performance Laboratory Project

Let's look at another challenging LEED Platinum project, a high-altitude [6300 feet (1910 meters)] laboratory, a joint venture project between Sierra Nevada College, the University of California, Davis, the Desert Research Institute, and the University of Nevada, Reno. The Tahoe Center for Environmental Sciences is a 45,000-square-feet lab and classroom building. Todd Lankenau was the principal in charge and project architect for the architectural firm, Collaborative Design Studio and Peter Rumsey was principal in charge and design engineer for the mechanical engineers, Rumsey Engineers.* Todd Lankenau talked about how they began:

> The client's original goal was to achieve LEED Silver. We started the project by going through a detailed interview process with various consultants to find the most appropriate people to include on the design team. We selected consultants for their technical and creative abilities as well as their commitments to be very personally involved with the project. Our goal was to create a very hands-on team of highly motivated, communicative, and personable individuals to foster excellence in the design process, which was of particular importance due to the large number of stakeholders from different institutions and the complex nature of the building and regulatory environment. As the project progressed, it became evident how important those initial choices were, since the success of this as well as any project is a result of the collective enthusiasm of the participants, and their ability to be leaders and champion the cause of designing an exceptional project.

> We developed a detailed project program with the owners and user groups and put into writing a very specific room-by-room requirement. This was critical as a reference throughout the design process to be sure all of the owners' requirements were met. We were fortunate to have owners who were very committed to and regularly involved in the design process, as well as being highly knowledgeable in the building type and sustainable design practices.

---

*Interview with Todd Lankenau, Collaborative Design Studio, February 2008, and with Peter Rumsey, Rumsey Engineers, April 2008.

We conducted numerous design charrettes that included the design team members and the owners, plus the various user groups and advisors from Lawrence Berkeley National Laboratory, Carnegie Mellon University, an architect from Germany, and others. The charettes were designed to optimize and integrate building systems while at the same time reduce energy consumption and initial construction cost. Our first goal was to reduce the heating and cooling loads of the building by optimizing the building envelope and reducing otherwise assumptive loads such as plug loads. We then studied various mechanical systems, and selected a hybridized system which resulted in a 100 percent outside air system with an energy savings of approximately 60 percent. The integration of these systems into a building, which had a predetermined footprint and severe height restrictions due to the regulatory agency requirements in the Lake Tahoe area, created a separate challenge in itself, requiring significant design creativity. It was also necessary that the exterior design of the building reflect the Lake Tahoe alpine architectural vernacular which further restricted the choices of style, form, and building materials.

There were essentially two types of charrettes and workshops generally categorized by size. The larger charrettes were important and provided the opportunity to step back and listen to divergent opinions and provided a forum for stimulating discussions while providing some long range perspective which is needed and helpful at times to prevent you from getting tunnel vision. However, the most productive and focused meetings were smaller group workshops of 8 to 12 people consisting of the core design team members with representation from the owners and the user groups. The smaller group size enabled us to focus on, and further evaluate, specific issues which may have been raised at the larger charrettes, but also to refine the details of the design concepts. It was optimum for our project to have the larger design charrettes at intervals of about 60 days, with focused smaller workshops at two week intervals.

There was a certain "bottom line" philosophy, according to Lankenau, about how to use LEED as a metric, considering that it is still, and will probably remain for the foreseeable future, a work in progress:

Our goal was this: if there was a choice between good design practice and sacrificing the best design solution in an effort to attain a LEED credit, we would always choose good design practice. As a design team, we agreed that we would try never to design something just to attain a LEED credit, but rather, if it appeared close, we would simply work harder to refine the design, and through the additional effort, perhaps achieve a better design which would result in achieving the credit. It can be a formidable challenge to avoid the temptation to accumulate additional credits, but I believe that good design practice will yield a well-designed building incorporating exemplary sustainable design principles which may or may not qualify as a LEED credit.

Of course, the temptation during the construction documents phase is always to get just a few more points, to get to the next certification level, a practice that inevitably distorts the goals of LEED and probably at this stage adds cost to the project without attendant gain. Lankenau says there has to be a balance and that at the construction document (CD) phase level it can be very effective to try harder:

Our philosophy was that if we're so close on so many of the credits, if we roll up our sleeves and refine the design, we can achieve something that's even better. Rather than being satisfied with let's say a 40 percent energy savings as compared to an ASHRAE 90.1 equivalent, we said, why not make it 50 percent. When we reached 50 percent, we said why not 60 percent? We ultimately ended up achieving a savings of about 63 percent. We did that just by continually refining elements of the design. An example of that is the ductwork. Every time there's an elbow in ductwork, it creates a static pressure drop, which makes the fan work a little bit harder, which in turn consumes additional power. [For example], we went through an exercise of reviewing the mechanical system and eliminating every conceivable elbow that wasn't necessary. We achieved our goal merely by going to that level of detail during the design process and rigidly enforcing it during construction. This was also coupled with continual review and improvement of the efficiency of the building envelope and further reduction of loads where possible. This is an example of taking what was an already a highly efficient and cost effective system and making it better. The mechanical and electrical engineers deserve significant credit for being so dedicated to the continual improvement of the building systems efficiency.

In the end, the Tahoe Center for Environmental Sciences received a LEED Platinum rating with 56 credits, which has delighted the owners who were only hoping for a LEED Silver rating. It also was awarded the Best Overall Sustainable Design Project for 2008 from the University of California System, among others, and has become a model for the design of energy efficient laboratory buildings and sustainable design practices.

PLATINUM PROJECT PROFILE

Tahoe Center for Environmental Sciences, Incline Village, Nevada

Completed in the fall of 2006, the Tahoe Center for Environmental Sciences houses classrooms and laboratories for programs that focus on understanding and protecting fragile alpine lakes. The three-story, 47,000-square-feet facility cost $25 million and is used by Sierra Nevada College, the University of California at Davis, the Desert Research Institute, and the University of Nevada, Reno. The facility uses 60 percent less energy and 30 percent less water than a comparable building. A variety of mechanical designs were used including chilled beams, displacement ventilation, radiant floor heating, overhead radiant heating and cooling panels, a turbine with cogeneration, lab exhaust heat recovery, 30-kW of building-integrated photovoltaics, nighttime chilled water production with a cooling tower and 50,000 gallons of chilled water storage, direct evaporative cooling in air handlers and a demonstration solar hot water heater. Water-free urinals and low-flow toilets were installed in the facility. A snowmelt/rainwater catchment system captures water for reuse.*

---

*Heather Livingston, "Tahoe Science Lab Goes for Platinum-LEED," *AIArchitect*, October 27, 2006. Kate Gawlik, "Active and Passive," *Eco-Structure*, November 2006.

Photography by Van Fox, courtesy of Collaborative Design Studio.

Peter Rumsey led the engineering team. He says that having the LEED Platinum goal helped even during the CD phase of the project, when some teams might think it's too late to make significant changes:

There was clarity in the goal. The owner said, "We're going to go for the highest LEED rating possible and if you can get Platinum that would be great." At the eleventh hour in the CD phase of the design, they said, "If we need to spend a little bit of extra money on a couple of the LEED features of the building, we're willing to do that." A couple of elements of the design were added then to push it over the top and get it into LEED Platinum. All of the design team members were on the same page, and were confident that we could do it.

This is the first lab building in the country to use chilled beams, which are a way of heating and cooling without using reheat [the energy-inefficient practice of cooling down outside air for general distribution and then reheating it for certain rooms]. It's one of only a small handful of the labs that don't use reheat for air-conditioning the lab, eliminating reheat can be a gigantic energy-saving measure. The chilled beams save on construction costs as well. I think not only is it a LEED Platinum building, but for a lab, it's a real breakthrough design. Since this building has been built, many labs are using or have seriously considered using chilled-beams.

GENERAL CONSTRUCTION ISSUES

During the CD phase, we have to write all the project specifications and get all the details on the drawings, so that the general contractor and various subcontractors can actually bid the project. These considerations naturally give rise to a number of important questions.

1 How will the project be bid, and how will this impact the inclusion of sustainability criteria in construction documents? Will it be a negotiated bid or a competitive bid?

2 Is there a CM/GC process (or CM at risk) in place, so that we can involve the general contractor or construction manager at this stage, to familiarize them with the specific sustainability features of this project?

3 Do we have an updated LEED scorecard ready to share with the team and to make part of any value engineering exercises?

4 Do the contract documents/specifications clearly reference the sustainability goals of the project? Have the goals and related actions for the contractor been incorporated into a "green" Division 1 specification section?

5 Will we use a checklist approach at coordination meetings, to follow through on our design intentions from the schematic design and design development phases?

6 Are we using LEED project management software to keep track of choices we make, or are considering, in the detailed design phase?

7 For any materials, systems or processes that may be interpreted by bidders as unconventional, has the design team researched the local/regional availability of the items and provided contacts for sources of additional information in the specifications?

8 Has the design been reviewed thoroughly to avoid areas of unnecessary overdesign and to incorporate integrated design systems such as green roofs for stormwater management, or daylighting for energy conservation?

9 Do we require the general contractor to provide an erosion and sedimentation control plan and documentation of compliance with the plan, even if the specific activities are required by code in the project city or county?

10 Do the construction documents make clear what documentation will be expected from the contractor to comply with certification and incentive programs?

11 Do the construction documents make clear that substitutions will be reviewed relative to the environmental goals of the project, as well as relative to more conventional criteria?

12 Have we reviewed all construction details to ensure that they use materials efficiently?

PLATINUM PROJECT PROFILE

Verdesian, New York, New York

A 27-story residential building in New York City's Battery Park City, the Verdesian was completed in 2006. The 300,000-square-feet luxury apartment building includes 252 residences; the total project cost was $75 million. A natural-gas-fired microturbine produces 70 kW of power (20 percent of the base load) and recaptures enough heat to provide 100 percent of the occupants' hot water demand. The Verdesian was designed to be 40 percent more energy efficient than a standard building. A nearby sewage treatment plant purifies wastewater for toilet and cooling tower use. Up to 10,000 gallons of rainwater can be harvested and used to irrigate the green roof garden. Three roof heliostats capture and redirect sunlight to a park below.*

QUESTIONS TO ASK DURING THIS PHASE

Let's consider now some typical questions that need to be asked at the construction documents phase for any high-performance/LEED project.

ENERGY ISSUES

Wherein this phase we need to get very specific about energy-using systems and components, as well as building commissioning.

1 What is the final level of energy efficiency we are planning to achieve? Is it possible to increase that level with more efficient equipment choices, without changing the overall design concept or appearance?

2 Has the commissioning authority clearly reviewed all systems documents and do we have a clear idea of the owner's project requirements (OPR) and the basis of design (BOD), as called for in the LEED building commissioning process?

3 Have we designed a measurement and verification plan and physical systems for the building consistent with Option B or Option D the U.S. Department of Energy's International Performance Measurement and Verification Protocol (IPMVP, 2003 edition) for active mechanical and electrical systems? If we are going to collect data on the energy efficiency of major building systems and components, who is going to be in charge of analyzing data?

4 Does the building glazing provide for alternative thermal and sunlight characteristics on each orientation? Are the façades designed so that they can be upgraded over time as fashions or conditions change? For example, can we add dynamic external or internal shades and shutters at some time in the future, if today's budget won't allow it?

5 Will the plumbing and wiring of the building allow for future environmental technologies as well as general changes in technology to be easily incorporated?

6 Did we provide for LEED-required basic and advanced commissioning support from the mechanical, electrical, and controls contractors in the specifications?

---

*Horizon Engineering Associates [online], http://www.horizon-engineering.com/hea_site/portfolio/residential/verdesian/verdesian.html, accessed April 2008. Meredith Taylor, "The First LEED Platinum Residential High-Rise: Batter Park City's Verdesian," Green Buildings NYC, January 17, 2008 [online], http://www.greenbuildingsnyc.com/2008/01/17/the-first-leed-platinum-residential-high-rise-battery-park-citys-verdesian/, accessed April 2008.

7 Has our commissioning agent or authority prepared a commissioning plan at this stage?

8 Do the specifications provide for an independent commissioning agent to review selected equipment submittals from the contractors? Who will be responsible for training building operators?

INDOOR ENVIRONMENTAL QUALITY

1 Does the HVAC design clearly separate areas of major chemical mixing or printing from other users in the building? Can we use green cleaning measures to get rid of chemical mixing entirely or avoid the need for separate ventilation systems?

2 Does the HVAC design provide for real-time monitoring of temperature and humidity in the building, with feedback to operational controls?

3 Have we clearly specified external shading systems or high-performance glazing components to support our daylighting design?

4 Can some of the external shading be designed to support building-integrated photovoltaics?

5 Does the architectural design provide for structural floor-to-floor partitions for areas of high pollutant mixing or generation, such as high-volume copying and printing rooms?

6 Is there any smoking allowed in the building (e.g., in a bar or restaurant)? If so, is there a separate ventilation system for such areas, with negative pressure, a separate return and deck to deck partitions?

7 Are we specifying carbon dioxide monitors to control ventilation levels in the building and especially in areas of high-occupancy, such as conference rooms?

8 Are we choosing carbon dioxide monitors for our project that have a long life in the field, an accuracy of at least 75 ppm and can go at least 5 years between calibrations?

9 If we have operable windows, how are the HVAC zones going to be controlled?

10 How will disputes between users over when to open the windows be mediated? Will we rely on an "impersonal" system such as red lights and green lights?

11 Have sensors for all lighting controls been specified for long life? Have they been located in areas that are easy to service, and can they be replaced easily?

12 Have we clearly specified the use of low-VOC paints and coatings in the building?

13 Have we clearly specified the use of low-VOC sealants and adhesives everywhere in the construction in the building?

14 Have we clearly specified the use of low-VOC carpets and floor coverings in the building?

15 Have we investigated how to prove that all engineered wood or agrifiber products in the building are free of urea-formaldehyde (UF)?

16 Will all casework be constructed of UF-free products? Will it meet low-VOC offgassing standards?

17 Have we designed permanent entryway systems (grills or grates) to capture dirt and contamination and keep it out of the building?

18 Have we required the mechanical contractor to comply with the Sheet Metal and Air Conditioning National Association's (SMACNA's) 1995 published standards for indoor air quality management for occupied spaces under construction, to protect absorptive materials from moisture damage and to keep ductwork clean and protected from contaminants?

19 Have we required the contractor to replace all filtration media prior to occupancy with MERV-8 filters? Have we sized the fans for high-efficiency MERV-13 filters and provided spaces in the duct system for thicker filters?

20 Does our building schedule allow for a 2-week flushing of the building at 100 percent outside air after completion and prior to occupancy? Is this a practical consideration in our region?

21 If this is not possible, do we have funds to require a baseline indoor air quality testing procedure that will meet U.S. Environmental Protection Agency protocols?

22 If the building will be occupied in phases, how will we accomplish the proper flushing of the building without exposing occupants to off-gassing of possibly noxious chemicals?

PLATINUM PROJECT PROFILE

The Willow School Art Barn, Gladstone, New Jersey

The Barn houses the school's middle grades, a cafeteria, the performing arts center and science center. The 13,000-square-feet facility cost $3.2 million. The building utilizes a wastewater management system where plants are grown hydroponically in sewage, which is treated and cleaned before it's returned to the ground. Harvested rainwater is used for toilet flushing, contributing to a 58 percent reduction in water use compared to a building built to code. A photovoltaic system provides 37 percent of the building's energy demand. The building was designed to use 70 percent less energy than a conventional building.*

WATER EFFICIENCY

Considering the growing importance of water conservation in buildings, at this time we need to check that the project has taken all possible steps in specifying components that will help us meet stringent water efficiency goals.

1 Have we specified water-conserving fixtures and sensors sufficient to achieve our water conservation goals for the project? Can we save 40 percent of water use against a baseline building, and qualify for a LEED Innovation point? Are all proposed systems and fixtures locally code-approved?

---

*Corinne de Palma, "A Barn for a Schoolhouse," Environmental Design & Construction, Marcy 3, 2008 [online], http://www.edcmag.com/Articles/Article_Rotation/BNP_GUID_9-5-2006_A_10000000000000275462, accessed April 2008.

2 If we're using harvested rainwater, does the specification require continuous labeling along the length of all piping?

3 Does UV treatment of harvested rainwater have an option for system shutdown in the event UV light bulbs fail?

MATERIALS ISSUES

The construction documents phase is our last chance to look once more at "environmentally preferable purchasing" policies for all the materials used in the building.

1 Is there an interest in making the building "vinyl-free" by specifying resilient flooring from other materials and finding alternatives to PVC pipe and conduit? Is this an important consideration for any of our stakeholders?

2 Have we specifically requested the use of salvaged or reclaimed building materials wherever possible? Do these materials account for at least 5 percent of the value of all building materials in the project? What will it take to reach that threshold?

3 Can we increase our goal for the use of salvaged or reclaimed materials to at least 10 percent of the value of all building materials in the project? What will it take to reach that threshold?

4 Does at least 10 percent of the value of all building materials come from post-consumer and/or post-industrial recycled content materials? Do we have ready access to this information? What will it take to reach that threshold? Could we get to 20 percent recycled-content materials (this is generally not that difficult in many urban areas).

5 Have we specifically requested that at least 10 percent of the value of the materials used in this project are extracted, processed and manufactured from local or regional sources (i.e., within 500 miles of the project)? Do we have ready access to this information? What will it take to reach that threshold?

6 Can we specify building materials so at least 20 percent of all materials (by value) are extracted, harvested, or recovered in the region? What will it take to reach that threshold?

7 Have we specified such items as bamboo flooring, linoleum, agricultural fiber MDF boards, so that rapidly renewable materials will make up at least 2.5 percent of the total value of all building materials used in the building? What will it take to reach that threshold?

8 Have we required that at least 50 percent of the value of all wood used in the building come from FSC-certified forests and have an acceptable chain-of-custody documentation? What will it take to reach that threshold?

9 Have we investigated local sources of supply for such items as FSC-certified wood to make sure they are available for this project?

10 Do our specifications require the general contractor to develop and implement a construction waste management plan? Are there opportunities locally for recycling construction waste if we comingle waste streams onsite?

11 Does this plan provide for a minimum of 75 percent recycling (by weight or volume) of all construction waste?

12 Is it possible to require the recycling of larger proportions of construction waste, such as 95 percent or more, and thereby qualify for a LEED innovation credit?

PLATINUM PROJECT PROFILE

The Christman Building, Lansing, Michigan

The Christman building is the first building to be certified at both LEED-CS Platinum and LEED-CI Platinum levels. A renovated 1928 landmark, the Christman Building is located on a former brownfield. Over 90 percent of the former walls, roof, floors and office furnishings were reused. The $12 million project is expected to save $40,000 a year in energy costs. The Christman Company, who provided construction management, historic preservation, LEED coordination and real estate development services for the project, occupies about half of the six-story, 62,000-square-feet class A office building.

Courtesy of Gene Meadows and The Christman Company.

LEED PROJECT MANAGEMENT ISSUES

Now is the time to begin preparing our LEED "design phase" submittal, and there are a number of questions that arise at this stage.

1 Are we fully prepared to submit our LEED design credits for certification review prior to beginning construction?

2 Have we determined which LEED innovation credits we are going to pursue, whether for exemplary performance or for achievements not presently covered by the LEED rating system?

3 Do we have systems in place to secure those innovation credits?

4 Have we investigated "borrowing" credits from other LEED systems, such as tenant build out guidelines, environmentally preferable purchasing policies, furniture and furnishings purchases, green cleaning and green exterior maintenance, and make them part of our operating policies going forward?

Bidding and Negotiation

Bidding and negotiation is an overlooked aspect of sustainable design projects; during this phase, building owners, developers and design teams have to work with contractors to achieve overall sustainable objectives, within the context of the project budget and the ease of construction. Often, in a public bid situation (more than a third of all LEED-NC projects are for public agencies), it's a good idea to devote a specific amount of time at "pre-bid" conferences to the LEED project requirements, to ensure that the bids will reflect the actual scope of work required from the contractor.

1 Have specifications and drawings provided enough detailed information to ensure that contractors will bid the work based on its specified systems, rather than applying a general "green premium" to the job, owing to uncertainties?

2 Is there a procedure for identifying and recruiting local builders, suppliers, and craftsmen that have experience and an interest in constructing green facilities?

3 Is there a procedure for informing potential bidders of the project's environmental priorities and goals (example: have we incorporated green program goals and explanation of the LEED rating system into the pre-bid conference for contractors)?

4 Have the contractor selection criteria included evaluation of past participation and performance in green construction projects?

5 Have we examined the bids to see that the sustainability programs and specific green building measures are included?

6 Is there a procedure that encourages the bidders and/or the selected contractor(s) to be a true participant in identifying alternative materials, systems, technologies, and/or methodologies to help meet the project's objectives during the construction process?

13

CONSTRUCTION AND OPERATIONS

This is a book about designing and delivering high-performance projects. Since most of the money in a project's life is spent in the construction and operations phases, those phases merit entire books on their own. For an integrated design process, the construction period is where "the rubber meets the road," a time when all of your grand intentions and careful design ideas have to be realized in the messy process of getting a building from a set of drawings and a hole in the ground to a finished project in which people can live, work, study, or play for decades to come.

Construction

I have found it useful on LEED projects to make sure that the construction process starts with a full explanation of the LEED goals and specifically which LEED credits the construction team is charged with achieving. If the construction documents were prepared properly, all of these requirements should have been in the General Conditions, Division 1 specifications. However, it's not generally the case that everyone reads all the specifications. That's why actively managing the construction process in high-performance projects is critical to achieving the desired results. The construction kickoff meeting is the place where all this starts, and a good general contractor will use a portion of that meeting to bring the LEED goals to the attention of the entire team.

John Pfeifer is senior vice president at McGough Construction in Minneapolis. His team recently completed (April 2008) an expected LEED Platinum project, a high-rise headquarters building in Minneapolis for Great River Energy. From the general contractor's viewpoint, he makes four key points about conducting a successful LEED Platinum project:*

1 Push the envelope on early decision making.
2 Intensify the early planning activity, before design even starts.

---

*Interview with John Pfeifer, March 2008.

3 Painfully, clearly assign which party is responsible for which credits and continually review that assignment. During the course of construction and toward the end when you want to submit your final report, there should be no question about what people have been doing. "We made assignments early on and kept revisiting those assignments because things change, people tend to get other ideas about who might be assigned to do what; that's just what happens in an 18-month long project. It's not an easy situation," Pfeifer said.

4 Without a committed owner, it isn't going to happen.

McGough Construction is a strong believer in a collaborative process model, much as other major general contractors are. Pfeifer believes that this delivery method is not only superior in general, but probably absolutely essential for the success of a high-performance building project. He says:

> There's a lot of hype these days about IPD, integrated project delivery. For years and years, McGough Construction in our region [upper Midwest] has pioneered the use of a collaborative or integrated delivery system. The vast majority of our projects for probably the past 20 years have been delivered with collaborative delivery systems. It has just proven over time to ensure the highest quality project with an efficient schedule at the lowest cost.

> Along with the collaborative delivery system is basically the open-book delivery mentality where all costs are subject to review and approval by the owner. We don't take the architect's plan, develop a bid, take it to the owner and say, "Take it or leave it." The owner is incredibly involved with all of this from day one—as is the architect and all of the other consultants—and he has continual buy-in every step of the way. At every step of the way, we're challenging numbers. Open book means that the owner sees all of the pros and cons, the pluses and minuses from the process that we're going through and becomes a big part of the decision-making process.

> With a high-performance and especially a high-level LEED certified project, you need to take many of those characteristics and crank up the intensity knob on all of them. You have to optimize the efficiency of the team working together. You have to intensify the planning efforts early on because so many of these things are interrelated to one another. The equation to put [a high-performance] building together is much more complicated than it was before. It's always a decision of weighing the benefits of one [approach] to another and comparing them with the costs.

Ted van der Linden is sustainability director at DPR Construction, a large California-based general contractor committed to a collaborative delivery process. He's been involved with delivering high-performance projects almost since the inception of the LEED rating system. Most of the general contractor's work in a LEED project is making sure that the subcontractors follow the plans and specifications of the design team. For DPR, it's mostly about educating their subcontractors and securing their cooperation.*

---

*Interview with Ted van der Linden, DPR Construction, February 2008.

On our projects we really focus on making certain that the subcontractors are very aware of why we're doing what we're doing. Obviously, they are instrumental in the delivery of a project. We don't want to throw them curve balls left and right, such as, "No, you can't use high-VOC paint. You can't use that sealant because it's got a high percentage of VOC's in it."

We've turned it around and really taken the lead on projects because [in many cases] we help write the specifications. We help the subcontractors get educated very quickly on what [LEED] means to them. We don't just give them the LEED binder and say, "Yep, we're targeting LEED Silver, good luck!" We make sure at a very early stage that they understand exactly what our intentions are from a green perspective. Their participation and awareness have changed greatly over time. In 2000, on some of our projects, it was literally like pulling teeth to get them to slow down, look at what we wanted to do and not price things ultra-conservatively, for example, because it was something different, it must cost more.

That was the big paradigm shift several years ago—the realization that a green building costs more in many situations. I always turned that around and said: "doesn't a better car cost more?" You can choose to buy a Yugo or you can choose to buy a Cadillac. One's going to cost more than the other because of its longevity, its quality and all of those things. Granted, a car is not necessarily the best analogy to bring to a building, but when expressed this way, people do get it.

More general contractors are becoming LEED-savvy, but many key subcontractors are still coming up to speed on LEED requirements. Here are some important questions to ask during the construction period.

1. Have we incorporated the green elements of the project and proposed certification into the kickoff meeting with all subcontractors who might be affected by them? Have we received agreement from the subcontractors to meet their obligations in helping us secure the desired LEED rating?
2. Are we assembling project LEED documentation as we go along? Is there a consultant or design team member specifically tasked with keeping all LEED documentation current and ready to submit?
3. Are we tracking the LEED points achieved (vs. our goals for various levels of certification) as we go along? Are we using LEED project management software to help with this task?
4. Have we received submittals for all green materials and specific systems used in the project and incorporated them into a notebook that can help with LEED certification, future building maintenance, and operator training?
5. Is the commissioning agent reviewing submittals for all systems to be commissioned, to determine compliance with design intent, per the requirements for enhanced or advanced commissioning (Fig. 13.1)?
6. Is the general contractor documenting compliance with the erosion/sedimentation control plan?
7. Is the general contractor or the waste management vendor documenting diversion of waste from landfills, at least at the 75 percent level? Do we have regular reporting of

Figure 13.1   The commissioning agent, Ken Urick of SSRCx Facilities Commissioning, is reviewing the chiller set-up with a site maintenance technician. *Courtesy of SSRCx Facilities Commissioning.*

progress toward this goal during construction? Is the waste management vendor supplying us with regular reports on waste diversion and recycling that we can use for LEED documentation?

8 Is building commissioning happening on the scheduled timetable?

9 Will the mechanical and electrical contractors be onsite during the functional testing period of commissioning to fix any problems found then?

10 Have we trained the maintenance staff in the specific details of operating all equipment?

11 Is the mechanical contractor tasked with and committed to following the SMACNA guidelines for maintaining indoor air quality during construction?

12 If we have made provisions for a 2-week building flush-out with 100 percent outside air, per the LEED protocol for verifying indoor air quality prior to occupancy, is that going to have any unforeseen schedule impacts?

13 If we choose an alternate pathway to demonstrate indoor air quality, for example, because local climatic conditions make a flush-out unwise, have we made arrangements to take samples of indoor air quality according to the LEED-approved protocols?

14 Have we documented this training, for future staff, and adequately enough for the enhanced commissioning LEED credit?

15 Have we prepared a systems and operations manual for this building, according to the LEED enhanced commissioning credit requirements?

16 Have we provided sensors and data collection for measurement and verification of key energy and water using systems?

17 Can we certify the building prior to occupancy, by having a commissioning contract in place? When do we have to finish all project documentation to accomplish this goal?

18 Are we going to finish the project on time, on budget and with the LEED rating we are expecting?

Without the talent and ingenuity of the general contractor, it's almost impossible to deliver high-performance buildings such as very large LEED Platinum projects (Fig. 13.2).

Michael Deane is manager of sustainable construction for Turner Construction Company, the nation's largest commercial building contractor. He speaks to the importance of the general contractor in the integrated design process in this way.*

> As contractors, we know a lot about constructability. We know a lot about which materials are low-emitting, have recycled content and are available locally, as well as what systems are energy- and water-efficient. We also know a lot about costs and

Empire State Building

Bank of America Tower (under construction)

Figure 13.2 **Imagine building the world's largest LEED-Platinum project in the crowded confines of New York's Times Square area, and you'll appreciate how much contractors bring to each LEED project certification.** © *Photography by Cook+Fox, courtesy of Cook+Fox Architects.*

*Interview with Michael Deane, Turner Construction Company, March 2008.

material availability. For example, we are working on a $100 million continuing care retirement community with 250 units. The project was 50 percent through construction and the owner came to us and said, "I was wondering if we could make this a LEED project." This is almost the worst possible scenario from a timing standpoint, because most decisions about sustainability had already been made and most of the materials and equipment had been purchased.

We looked at what was in the design and what could still be procured and the materials that hadn't already been bought. For instance, they specified 1.6-gallon flush toilets and they were already ordered and sitting in a warehouse somewhere so there was no way they were going to re-spec them. That was a missed opportunity. If we had been sitting at the table during design, we would have said, "You know, there are about 120 toilet fixtures out there that have lower flush rates. I think we can find one within your budget, that you will accept from an aesthetics view point and you would have the opportunity to get a water efficiency credit." So that is an example of an opportunity lost.

On the plus side, we were able to re-order the drywall from a plant that produced 100 percent recycled material with no change in price or delivery schedule and with that single purchase, we were able to achieve both a local/region material credit and a recycled content credit. So that was an opportunity gained at the last minute.

For Deane, the biggest challenge in integrated design is to get the owner and architect to engage the contractor early in the process and to include them as a full participant in the design team, even on small projects. On very large projects, contractors are almost always engaged early on (but it doesn't always mean that designers are listening to their advice.)

Once an architect asked me, "What do you want from us?" I said, "I want you to listen to us." The challenge [as a contractor] is to be recognized as an equal player and to get a seat at the table. Then the challenge is one of education because we have found that the single biggest driver of increased costs in green building is lack of knowledge. This stuff isn't hard; it's just that not everybody knows it yet. We spend a lot of time informing people about what the realities are—the realities of cost, product availability, installation difficulty and economic payback. We live with that day in and day out because we're at the end of the chain. An architect can design a beautiful building but at the end of the day, the owner is going to want it delivered on time and on budget, and that's our job.

Paul Stoller of Atelier Ten spoke about the importance of the construction team on Yale University's Sculpture Building and Gallery, a project I've cited before in this book.*

Shawmut, the construction mangers, were superstars. When you go into construction, there are a fair number of [LEED] credits that come down to the construction manager being on the ball and making sure that submittals come through that meet the

---

*Interview with Paul Stoller, Atelier Ten, March 2008.

standards expected. Shawmut was fantastic at holding all of the subcontractors up to the standards that were set out in the design documents and specifications. Shawmut's careful attention to product sourcing and performance standards as well as their construction standards and methods were critical. Had they not paid such careful attention to the environmental aspects of the project, we would not have made it to Platinum. *That confirms lessons we've learned on other projects: when you have an attentive, dedicated construction manager, achieving a high-performance building is much easier.* If you've got an inattentive or dismissive construction manager, it's a nightmare.

Occupancy and Operations

Too often overlooked in the daunting task of getting a project built is how it will be operated over its lifetime. Recent USGBC studies demonstrate that LEED-certified buildings are in fact getting about 30 percent average energy savings as predicted by computer models. However, the studies also found a wide variation in individual outcomes, suggesting that individual buildings operations are critical in achieving desired results.* Here are a few key questions you need to ask that will optimize the environmental benefits of building operations in the future.

1 Will end-users be fully oriented to the systems when they take over the building, and are there provisions for an operating manual and/or interpretive information to help future users learn about the building design?

2 Is there a contract in place for near-warranty-end review of the commissioning of all key mechanical and electrical systems, along with water-using systems?

3 Have we committed to a periodic survey of occupant satisfaction with thermal comfort during the first 6 to 18 months of occupancy, especially one called for in the LEED rating system, and are we committed to making adjustments as needed to increase satisfaction levels?

4 How will maintenance staff be notified and instructed in future years to check calibrations on all HVAC, carbon dioxide and lighting sensors?

5 Can the energy performance and the lighting systems be monitored from the engineer's office for ease in troubleshooting the systems?

6 Have we begun collecting energy and water use data from the measurement and verification system, for comparison with projected (modeled) energy and water use?

7 Have we incorporated educational objectives into the project through signage, public relations, brochures, tours, seminars, and other means?

8 Is there a commitment and a budget to begin the process of LEED-EB certification within the next 2 years, to see how our long-term operations are affecting energy use and environmental quality metrics?

9 Can we begin collecting data now that will assist in future LEED-EB certification, such as commuting patterns and recycling rates?

---

*New Buildings Institute, study for the USGBC, 2008, www.usgbc.org/docs/NBI%20and%20Group%20Release%20040108.pef, accessed April 1, 2008.

10 Do special inspection programs (such as commissioning) include ongoing review of critical functional elements?

11 If this is a school or college project, have we trained the teachers or instructors to explain and use the sustainability systems of this project?

12 Have we explained the benefits of the sustainable design features in this project to higher-level decision makers, to get their support for sustainable operations?

13 Have we incorporated LEED-EB requirements such as environmentally preferable purchasing, transportation demand management, and enhanced waste management policies, into our ongoing building operating plans?

LEED for Existing Buildings Operations and Maintenance (LEED-EB) is the system for benchmarking, assessing, and certifying on-going operations. Consideration of the role of LEED-EB in operating and maintaining high-performance buildings is beyond the scope of this book. However, it's worth noting that the world's largest property management company, CB Richard Ellis (CBRE), has made a major commitment to the LEED-EB certification process. Sally Wilson is senior vice president and global director of environmental strategy at CBRE. Here's her approach to LEED-EB.*

We made a commitment with the U.S. Green Building Council in November [of 2007] to register 100 of our buildings in the LEED-EB portfolio program. Right now we're going through the process of selecting and registering the buildings, and we're probably going to have closer to 150 buildings. We will have significant number of them certified by the end of 2008.

On the other hand, we still have continued to develop our "Sensible Sustainability" solutions within our asset services division that deals with energy use, water use, waste management and expanded recycling programs for light bulbs, batteries and printer cartridges. We're really trying to extend it and make the recycling programs better. One of the big components that we're going to work in this year, now that we have the infrastructure in the buildings, is communicating with the tenants to help them understand how they can change their behavior to act more responsibly and utilize the programs that we set up. It really doesn't matter how energy efficient we've made the building or whether we have these waste programs in place. It's the tenants in the buildings that create the use primarily. We have to communicate to them so they can take advantage of what we've done and ultimately make the reduction through their actions.

One of the key elements in operating a building for the health and productivity of the occupants is a green cleaning program. For Wilson, green cleaning is an important part of the CBRE program.

On the cleaning side, we put solutions in place to ensure that the janitorial services are turning lights and equipment off at night. The other thing with janitorial services

---

*Interview with Sally Wilson, CBRE, February 2008.

is we help them put green practices into place, for example cleaning with green products. A lot of it is just awareness. Once people realize what they're doing, they realize their impact.

It's not that difficult to get a green cleaning program in place. Frankly, the service providers really welcome the change that's happening on the cleaning side because it's healthier for them. When CBRE is managing the building, it's usually not an issue to go to the janitorial company and ask them to make these changes. I'm a tenant broker and when I go to a building that's not a CBRE-managed building, I generally get some resistance, but once you push them further and make them dig a little deeper they realize that the janitorial service won't really have a problem with it. In a sophisticated market most of the janitorial services are already going to have a [green cleaning] program.

Over the long run, sustainable operations are absolutely critical for reducing carbon footprint. While I have focused in this book on the integrated design process for new construction, LEED-EB and other methods of encouraging building owners and tenants to change behavior and reduce the environmental impacts of existing buildings are very important for the future of global warming, as well as improving the health and productivity of both people and planet.

14

LOOKING AHEAD—DESIGNING LIVING BUILDINGS

Along with many others, Bill Reed writes about the need to continue the push for high-performance buildings into the realm of restorative and regenerative design. In a 2005 paper, Reed (and his coauthors) wrote: "the term 'regenerative' is useful because it suggests the self-organizing, self-healing and self-evolving properties of living systems."[†] Rather than green design (as LEED defines it), where the goal is just to reduce the damage by being "less bad," and even fully "sustainable" design, where the goal is to evaluate our impacts against a goal of "zero harm," the spirit of regenerative design envisions a trajectory of responsible design (Fig. 14.1) that aims to restore living systems to a more productive level than we found them, all the while maintaining a prosperous and healthy human existence. Quite a tall order!

One step along the way to fully regenerative design is to create a "living building."[*] Jason McLennan, currently CEO of the Cascadia Chapter of the U.S. Green Building Council, is a serious advocate for this concept. In a brilliant twist on the LEED-NC system, with its seven prerequisites and 69 credit points, often criticized for allowing buildings to be certified with only marginally better energy performance, for example, McLennan postulated a rating system, the Living Building Challenge (LBC), that has only prerequisites and no credits. In other words, you either have a living building or you don't.[‡] (Certain accommodations are made in the current version 1.2 of the LBC for present market realities.)

Here's how it works: there are 16 categories, and a project must fulfill all of the requirements to be certified as a Living Building. The system is performance based,

---

[*]Bill Reed strongly prefers the term "living systems," arguing that buildings can't by themselves be living. Nonetheless, the term "living buildings" has become descriptive of the next evolution of green buildings, so I'll stick with it.

[†]Bill Reed, Joel Ann Todd, and Nadav Malin, "Expanding Our Approach to Sustainable Design—An Invitation," Brattleboro, Vermont: Building Green, Inc., December 15, 2005.

[‡]Cascadia Green Building Council, www.cascadiagbc.org/lbc, accessed April 29, 2008.

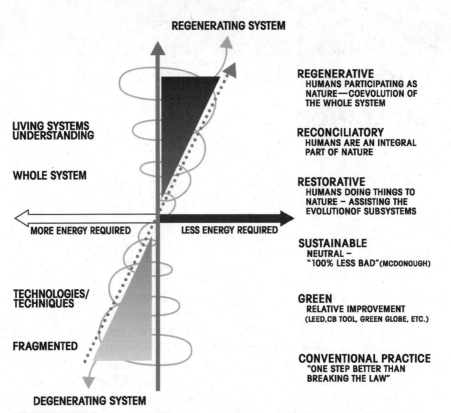

REGENERATING SYSTEM

REGENERATIVE
HUMANS PARTICIPATING AS
NATURE—COEVOLUTION OF
THE WHOLE SYSTEM

RECONCILIATORY
HUMANS ARE AN INTEGRAL
PART OF NATURE

RESTORATIVE
HUMANS DOING THINGS TO
NATURE – ASSISTING THE
EVOLUTIONOF SUBSYSTEMS

SUSTAINABLE
NEUTRAL –
"100% LESS BAD"(MCDONOUGH)

GREEN
RELATIVE IMPROVEMENT
(LEED,CB TOOL, GREEN GLOBE, ETC.)

CONVENTIONAL PRACTICE
"ONE STEP BETTER THAN
BREAKING THE LAW"

LIVING SYSTEMS
UNDERSTANDING

WHOLE SYSTEM

MORE ENERGY REQUIRED   LESS ENERGY REQUIRED

TECHNOLOGIES/
TECHNIQUES

FRAGMENTED

DEGENERATING SYSTEM

Figure 14.1 **Trajectory of environmentally responsible design,
showing a positive upward movement from conventional design through
green design to fully restorative and regenerative design.** © *Integrative Design
Collaborative and Regenesis, Bill Reed, 2006.*

so that "best practices" don't have to be referenced, as in LEED. To get to the results
for certification, you'll have to use them. Two of the categories relate to "beauty and
inspiration," underlying the reality that "ugly" buildings are not really sustainable
because they won't engage the human spirit and no one will want to keep them around.
You can understand the value of this viewpoint by considering how strongly certain
historical buildings engage the emotions of a community, even though they might be
"energy hogs" or otherwise not meet current LEED standards. We value them because
of the cultural context they create and express.

Recalling the discussion of BHAGs in Chap. 3, you'll appreciate the value of the
Living Building Challenge as setting "Big Hairy Audacious Goals" for a new generation
of living buildings and a new generation of architects, designers, and building owners.

In the Living Building Challenge evaluation system, the required categories are the
following:

1 Responsible site selection (this models LEED's Sustainable Sites Credit 1)
2 Limits to growth (don't build on any undeveloped site)
3 Habitat exchange (set aside land equal to the development area as habitat)

4 Zero net energy (the building should generate all of its energy from renewable sources, on a net annual basis)

5 No use of the 13 persistent, toxic, or bioaccumulative materials (such as formaldehyde, PVC, HCFCs, and the like)

6 Eliminate carbon footprint of building materials (offset carbon impact through purchase of offsets)

7 Responsible production and procurement of materials (only use FSC-certified or salvaged wood materials)

8 Appropriate materials and services radius (limits on how far materials can travel to the project site)

9 Construction waste recycling (requires at least 80 percent waste diversion of all materials)

10 Net-zero water use (except for potable water required for health systems, the building should only use captured or site-treated water)

11 Sustainable water discharge (handle 100 percent of stormwater and building water discharge onsite)

12 Civilized work environment (buildings must have operable windows with fresh air and daylight)

13 Source control of indoor pollutants (essentially following LEED Indoor Environmental Quality credits 4 and 5)

14 Healthy ventilation for indoor air quality (Exceed "code" based ventilation standards by 30 percent)

15 Design for spirit (this is worth quoting: "The project must contain design features intended solely for human delight and the celebration of culture, spirit and place appropriate to the function of the building.")

16 Design for inspiration and education (the building should provide educational materials to the public and be open at least once a year for tours)

There is a promised Living Building User's Guide to assist prospective projects in meeting these requirements in greater detail. Nonetheless, you can see that the Living Building Challenge represents a significant evolutionary advance in sustainable building thinking.

Hard Bargain Farm, Accokeek, Maryland

Let's take a look now at one project that has been designed to meet these standards, the Alice Ferguson Foundation's "Hard Bargain Farm," an environmental education facility located in Maryland, near Washington, DC. This project won the "Demonstrated Leadership" award at the 2007 USGBC "Greenbuild" show, for unbuilt projects.* The design concepts include the following strategies:

---

*Cascadia Green Building Council, www.cascadiagbc.org/lbc/lbc-competition/, accessed April 29, 2008.

1 Maintaining a compact development footprint that takes advantage of existing infrastructure.
2 Building on previously developed and degraded sites.
3 Nestling the buildings within and around mature trees.
4 Elevating the buildings to respect natural grade changes and minimize site disturbance.
5 Designing from the "inside-out," with building performance (daylight, energy efficiency, ventilation, water flows) guiding building form.
6 Movable insulation panels and sun shading devices that allow the buildings to adapt both daily and seasonally, to provide comfort with the least amount of energy consumption.
7 Use of wood grown and milled at Hard Bargain Farm.
8 Use of unfinished wood and/or metals that will weather naturally ("patina") and also facilitate eventual recycling.
9 Use of straw grown onsite for straw-bale walls of the one building.
10 Capture rainwater for use in the structures or on the site.
11 Provide a "living wall" to filter and channel roof runoff.
12 Use composting toilets or constructed wetlands for waste treatment.
13 Use salvaged or recycled "trash" materials for construction.

Philadelphia-based architect Muscoe Martin of M2 Architecture is a long-time proponent of sustainable design, who worked as the local architect for the University of Pennsylvania's Morris Arboretum project, profiled earlier in this book. Here he describes how a longer charrette process led to creative problem solving for an unusually challenging design program.*

Hard Bargain Farm is an environmental educational center designed jointly with M2 Architecture and Re:Vision Architecture, also located in Philadelphia. M2 Architecture is involved in the early design phases and then Re:Vision will pick up the project as it moves into construction.

We had a charrette that was different from the Morris Arboretum charrette. It was a four- or five-day event. We had staff members that worked for the foundation, board members who will be helping raise the money for the project, neighbors and interested community members. We chose to really make it a true design charrette in the sense that we actually developed the conceptual design for the project during the course of those five days. We were getting into it "real time" as we were developing the drawings. Of course, we had our engineers, landscape architects and our fully integrated design team there as well as the client's group. There were a couple of difficult decisions that had to be worked out in charrette. One was where to put the building on the site. They have quite a large site but there were some constraints. They wanted the building to be a living building. In fact, this project was one of the winners in the Living Building Challenge competition at Greenbuild 2007.

---

*Interview with Muscoe Martin, M2 Architecture, March 2008.

During the charrette we had to figure out where to locate the building, and one of the criteria established before the charrette was that it had to be a zero net energy structure. In that climate, this requirement means we would need to collect quite a bit of solar energy both for daylighting and heating. The ideal site for that would to place the building out in the sun with good solar orientation. But the client's preferred site was on the location of an existing building that they wanted to either expand or replace.

The problem: that site was in the shade. It was tucked into some beautiful, mature shade trees with very poor solar orientation. That was a real conflict that we had: being able to deliver a really energy-efficient building while meeting the client's desire to use land that was previously disturbed. We kind of banged our heads against this for a while in the charrette and we got to point where we were at a standstill. We took a break for lunch. My joint venture partner from Re:Vision Architecture, Scott Kelly, had gone outside with his lunch. We held the charrette on the project site and he was sitting in the shade next to the existing building that the client wanted to renovate. He looked down between his feet where he was sitting and he saw that there was a bunch of moss growing there. He started thinking about that. It was a shady area, but stuff was growing there. So he thought, "Maybe our building needs to be more like a moss and less like something that's out in the sun." He wasn't sure what that meant but it was just kind of an inspiration that came to him.

After lunch he shared that with the group and the proverbial light bulb came on in my mind. I said, "Of course! We need two buildings. We need to split up the building and have one set of uses on the existing site. It's going to be the 'moss' building, and it's going to collect water, which is what moss does. It's going to be a shady building and take advantage of that environment. It can have big windows without worrying about overhangs because it's got shade. The other building should be out in the sunny field. That's going to be used for daytime uses and it's going to collect solar energy. It's going to be like grass. It's going to be the 'grass' building. (Fig. 14.2)"

That was huge. Within that half-hour of discussion, we had solved the problem. It was because we had first banged our heads around this issue, then went away and let things settle. Then someone had an inspiration in which they brought back and created another set of inspirations amongst the group. That project is now moving along, we're in the design process and we now have two buildings. One will collect and purify rainwater, which will be distributed to both buildings. That's the Moss Lodge. The Grass Building will sit down in the field. It is smaller but it will have big eaves, big overhangs and photovoltaics. It will collect enough solar energy for both buildings. The two buildings will work together symbiotically.

There were other positive benefits of breaking the building up in terms of the uses. By having the client's staff in the room for the charrette, they were able to determine how the uses for each building should be divided. We had input from both the client, who knew how things were going to operate, and expertise from the design team, who knew how to incorporate energy efficiency, water conservation and water efficiency into the design.

This made us examine the idea of a living building. A building can't really live on its own because it's got to take in energy, materials and water and put out waste. You

Figure 14.2   The Alice Ferguson Foundation Hard Bargain Farm in Accokeek, Maryland will consist of two buildings: the "Moss Lodge" situated on a shady site and the "Grass Building" situated on a sunny site. Available resources and energy will be shared between the two to achieve the goals of a living building. The project expects to begin construction in 2009. *Courtesy of M2 Architecture/Re:Vision Architecture.*

can't think of a building as living, *you have to think of a site and the building together as a living thing*. Having two buildings, each of which are sitting in very different microclimates but within a few hundred yards of each other, means that they are going to have very different flows of energy through them. Each is able to make use of what it can do well and share that [with the other]. It's pretty unique. I hadn't encountered a situation like this before where we let two buildings work together in that way.

The two buildings will work symbiotically, with one supplying energy and the other water. The new "Moss Lodge" will replace the old overnight Wareham Lodge, located on a shady hillside. Its roof reaches up to gather rain that will be purified for use by both buildings. The landscape will channel and filter stormwater runoff and graywater from sinks and showers to recharge the underground aquifers. The day-use education center, called the Grass Building, will be built at the sunny edge of a field, its roof spreading out like wings to collect solar energy for the entire complex. The Grass Building will provide multi-functional space, indoors and covered outdoors, for students visiting Hard Bargain Farm.

Although small, this project is a wonderful example of the trend in restorative and regenerative design. It takes an understanding client, the right design program, an experienced design team and a process that allows for insight and innovation to make all of this happen. But, given that the size of the U.S. commercial and institutional building market, even in a bad year, exceeds $200 billion, don't we have enough money (and time) to "do the right thing," instead of just "doing things right"? In the developed world, we have the design talent, builders who can build just about any design, product manufacturers who innovate constantly, abundant capital, financiers who can finance just about anything, all the right stuff. Why can't we build high-performance living buildings—in all of our market sectors—as our legacy to future generations?

If you've read this far, you know that the answer is up to you and your colleagues. This book has made the case for an integrated design process, and it has shown you many examples of how it's done. Now it's your turn; good luck!

INTEGRATED DESIGN RESOURCES

American Institute of Architects, 2007, "Integrated Project Delivery—A Guide," available at www.aia.org/ipdg, accessed June 30, 2008.

Better Bricks, 2007, "Integrated Design Meets the Real World: Mount Angel Integrated Design Roundtable Discussion," three parts, available at www.betterbricks.com/DetailPage.aspx?ID=915, accessed June 30, 2008.

Busby Perkins+Will and Stantec Consulting, 2005, "Roadmap for the Integrated Design Process," British Columbia Green Building Roundtable, available at www.mucs.ca/library/G1%201%20IDP%20Roadmap%20Part%20Two.pdf, accessed June 30, 2008.

Good, Nathan, n.d., "What's an Eco-Charrette?" available at www.betterbricks.com/DetailPage.aspx?ID=275, accessed June 30, 2008.

Kwok, Alison G., and Walter T. Grondzik, 2007, *The Green Studio Handbook: Environmental Strategies for Schematic Design*, Amsterdam: Elsevier/Architectural Press, Chapter 3.

Malin, Nadav, 2004, "Integrated Design," *Environmental Building News*, November, available at www.buildinggreen.com/auth/article.cfm/2004/11/1/Integrated-Design, accessed July 31, 2008.

Prowler, Don, and Stephanie Viera, 2007, "Whole Building Design," in *Whole Building Design Guide*, available at www.wbdg.org/wbdg_approach.php, accessed June 30, 2008

Whole Building Design Guide, 2006, "Engage the Integrated Design Process," WBDG Aesthetics Subcommittee, available at www.wbdg.org/design/engage_process.php?ce=id, accessed June 30, 2008.

Whole Systems Integrated Process (WSIP) Guide 2007 for Sustainable Buildings & Communities, ANSI/MTS Standard WSIP 2007, available at http://webstore.ansi.org/RecordDetail.aspx?sku=ANSI%2fMTS+1.0+WSIP+Guide-2007, accessed June 30, 2008.

Yudelson, Jerry, 2007a, *Green Building A to Z,* Gabriola Island, B.C.: New Society Publishers, 93–95.

Yudelson, Jerry, 2007b, *The Green Building Revolution,* Washington, D.C.: Island Press, 168–173.

Zimmerman, Alex, n.d., "Integrated Design Process Guide," Canada Mortgage and Housing Corporation, available at www.waterfrontoronto.ca/dbdocs//4561b14 aaf4b0.pdf?PHPSESSID=320a40062358e082e363e50efd67b16e, accessed June 30, 2008.

INDEX

Note: Page numbers referencing figures are italicized and followed by an "*f*"; page numbers referencing tables are italicized and followed by a "*t*".